George Oliver

The Symbol of Glory

Shewing the Object and End of Free Masonry

George Oliver

The Symbol of Glory
Shewing the Object and End of Free Masonry

ISBN/EAN: 9783337396404

Printed in Europe, USA, Canada, Australia, Japan

Cover: Foto ©berggeist007 / pixelio.de

More available books at **www.hansebooks.com**

THE

SYMBOL OF GLORY:

SHEWING

THE OBJECT AND END OF

FREEMASONRY.

BY GEORGE OLIVER, D.D.,

RECTOR OF SOUTH HYKEHAM; VICAR OF SCOPWICK; PAST DEPUTY GRAND MASTER OF THE GRAND LODGE OF MASSACHUSETTS; PAST D. P. G. M. FOR LINCOLNSHIRE; HONORARY MEMBER OF MANY LODGES AND LEARNED SOCIETIES.

Sic itur ad astra.—VIRGIL.

NEW YORK:
MASONIC PUBLISHING AND MANUFACTURING CO.,
432 BROOME STREET.
1870.

CONTENTS.

VALEDICTORY ADDRESS;

Containing a personal narrative of the motives which induced the Author to become an expositor of Masonry.

DEDICATED TO THE FOLLOWING

Subscribing Grand Officers.

Bro. The Earl of Aboyne, P.G.M. for Northamps. and Hunts.
.. W. H. Adams, Esq. P.M., 339, Boston, P.G.S.W. Lincolnshire
The Hon. G. C. Anderson, P.G.M. Bahamas
.. Rev. E. Brine, B.A., P.G. Chap. Worcestershire
.. E. F. Broadbent, Esq. P.M. 374, P.G. Trea. Lincolnshire
. E. A. Bromehead, Esq. P.M. 374, and P.G.J.W. Lincolnshire
.. W. Browne, Esq. 407, Ireland, P.G.J.W.
.. W. Buckle, Esq. 707, Handsworth, P.G. Sup. Works, Staffordshire
.. John Burrows, P.G. Sw. B. Worcestershire
.. Rev. W. J. Carver, Representative from the G. L. of Massachusetts
.. J. D. Cherry, Esq. J.D. 69, Londonderry, P.G. Purs.
.. Rev. E. M. Clarke, J.W. 69, Londonderry, P.G. Chap.
.. Rev. J. E. Cox, M.A., Grand Chaplain
.. Sir Charles Douglas, M.P. 356, Warwick
.. Maxwell Daring, Esq. S.W. 640, Ireland, P.G. Purs.
.. Thos. Ewart, Esq. P.M. 646, D.P.G.M. Northamps. and Hunts.
.. Michael Furnell, Esq. S.G.I.G. 33°. P.G.M. North Munster
.. F. Foster, Esq. 69, Londonderry, P.G.S.W. Derry
.. Alex. Grant, Esq. D.P.G.M. Derry and Donegal; Hon. Member of 126, 265, 279, 282, and 284, England; and 46, 196, 332, 407, and 589, Ireland
.. T. D. Harrington, Esq. Z. Victoria Chap. D.P.G.M. Montreal
.. T. Hewitt, Esq. P.G.S.W. North Munster
.. J. A. Hicken, Esq. K.T.P.G. Stew. Madras
.. Rev. W. N. Jepson, W.M. 374, P.G. Reg. Lincolnshire
.. J. F. Johnston, Esq. 69, P.G. Sup. Works, Derry
.. J. G. Lawrence, J.W. 326, P.G.J.W. Madras
.. J. S. Langwith, Esq. P.M. 466, Grantham, P.G. Sup. Works, Lincolnshire
.. William Lloyd, Esq. P.M. 51, 135, 689, 707, & P.G. Trea Warwickshire and Staffordshire

Bro. Very Rev. Archdeacon Mant, P.G.M. Belfast and North Down
.. Jas. M'Murray, Esq. J.W. 69, P.G. Sec. Derry
.. J. Maskell, P.M. 340, P. Dep. G. Sec. Madras
.. J. Middleton, P.M. 374, P.G. Sec. Lincolnshire
.. The Baron von Nettlebladt, P.G.M. Rostock, N. Germany
.. Rev. C. Nairne, P.M. 374, D.P.G.M. Lincolnshire
.. W. A. Nicholson, Esq. P.M. 374, P.P.G. Sup. W. Lincolnshire
.. G. Oliver, Esq. P.P.G. Stew. Lincolnshire
.. E. G. Papell, Esq. W.M. 326, P.G.J.W. Madras
.. J. W. Pashley, Esq. P.M. 611, P.P.G. Trea. Lincolnshire
.. Rev. T. Pedley, 646, Peterborough, P.G. Chap. Northamptonshire and Hunts.
.. G. W. K. Potter, Esq. P.M. 109, London, G.J.D.
.. J. Price, Esq. 69, P.G. Sec. B. Derry and Donegal
.. S. Rawson, Esq. P.G.M. China
.. W. Roden, Esq. M.D., P.M. 523 and 730, Kidderminster, D.P. G.M. Worcestershire
.. S. Rankin, Esq. 196, P.G. Stew. Derry
.. Rev. J. C. Ridley, P.G.M. Oxfordshire
.. H. Ridley, Esq. 69, P.G. Trea. Derry
.. W. Stuart, Esq. P.G.M. Hertfordshire, *(3 copies)*
.. R. J. Spiers, Esq. P.M. 425, Grand Sword Bearer
.. Capt. J. Stirling, 69, P.G. Stew. Derry
.. Augustus Tilden, P.G. Sup. Works, Worcestershire
.. W. Tucker, Esq. P.G.M. Dorsetshire
.. R. Turner, Esq. P.M. 466, Grantham, P.P.G.J.W. Lincolnshire
.. R. Taylor, P.M. 374, and P.G.D.C. Lincolnshire
.. Henry C. Vernon, Esq. P.G.M. Worcestershire
.. J. Wyld, Esq. M.P. Representative from the G.L. of Texas
.. E. B. Webb, Esq. 69, P.G. Stew. Derry

TO THE FOLLOWING

Subscribing Officers and Members of Private Lodges.

Bro. W. T. Adrian, P.M. Manchester Lodge, 209, Lambeth
.. R. J. Alexander, Light of the North Lodge, 69, Londonderry
.. E. W. Allpress, Bank of England Lodge, 329, London
.. J. Arnold, J.W. First Lodge of Light, 689, Birmingham
.. Algernon Attwood, P.M. Universal Lodge, 212, London
.. George Addison, Preston
.. Marcus Attwood, Universal Lodge, 212, London
.. J. N. Bainbridge, M.D., P.M. Bank of England Lodge, 329, London
.. T. Baker, I.G. Hope and Charity Lodge, 523, Kidderminster
.. Alexander Bolton, Dublin, *(6 copies)*
.. W. Boyd, P.S.W. and Sec. Social Friendship Lodge, 326, Madras
.. G. F. Brady, M.D. Light of the North Lodge, 69, Londonderry
.. S. Brizzi, P.M. Bank of England Lodge, 329, London
.. William Brooks, St. Thomas' Lodge, 166, London
.. C. F. Brown, J.W. Social Friendship Lodge, 326, Madras

Bro. J. Bull, Bank of England Lodge, 329, London
.. C. Cameron, Calcutta
.. J. E. Carpenter, W.M. Shakspere Lodge, 356, Warwick
.. G. Chance, Sec. Bank of England Lodge, 329, London
.. L. Chandler, P.M. St. Paul's Lodge, 229, Past Grand Steward, London
.. W. Clarke, P.M. Social Friendship Lodge, 326, Madras
.. R. Cobbet, Bank of England Lodge, 329, London
.. Fillippo Coletti, Bank of England Lodge, 329, London
.. F. Cook, Boston
.. C. H. Cornwall, J.W. Yarborough Lodge, 812, London
.. R. Costa, S.W. Bank of England Lodge, 329, London
.. M. Costa, J.W. Bank of England Lodge, 329, London
.. Rev. S. O. Cox, Light of the North Lodge, 69, Londonderry
.. J. T. Darvill, P.M. St. John's Lodge, 107, London
.. S. Dillet, P.M. Royal Victoria Lodge, 649, Nassau, Bahamas
.. A. Dimoline, W.M. Royal Clarence Lodge, 81, Bristol
.. W. W. Douglas, M.A. 730, Oxford
.. E. Dorling, P.M. Perfect Friendship Lodge, 552, Ipswich
.. J. Durance, jun. Witham Lodge, 374, Lincoln
.. Lieut. Connolly Dysart, Madras Army, First Lodge of Light, 69, Londonderry
.. F. Elkington, S.W. First Lodge of Light, 689, Birmingham
.. C. W. Elkington, P.M. & Sec. First Lodge of Light, 689, Birmingham
.. Capt. T. E. Ethersey, P.M. Universal Charity Lodge, 340, Madras
.. Lieut. Ethersey, P.M. Perfect Friendship Lodge, 522, Ipswich
.. W. T. Ethersey, P.M, Universal Charity Lodge, 326, Madras
.. W. Evans, P.M. Temple Lodge, 118, London
.. Rev. T. B. Ferris, Chap. Bank of England Lodge, 329, London
.. W. H. Fletcher, S.W. Hope and Charity Lodge, 523, Kidderminster
.. J. W. Foulkes, I.G. Bank of England Lodge, 329, London
.. — Frith, M.D. Calcutta
.. Hector Gavin, Edinburgh
.. E. Castellan Giampietro, Bank of England Lodge, 329, London
.. W. Gillman, D.C. First Lodge of Light, 689, Birmingham
.. Alfred Glover, S.D. 803, Longton, Staffordshire
.. R. Graves, P.M. Bank of England Lodge, 329, London
.. Wm. Green, J.D. 803, Longton, Staffordshire
.. W. B. A. Greenlaw, W.M. Lodge of Sincerity, 224, Plymouth
.. B. Hall, Trea. First Lodge of Light, 689, Birmingham
.. J. Hamilton, Esq. High Sheriff, Donegal, Light of the North Lodge, 69, Londonderry
.. John Harris, P.M. Albion Lodge, 9, London
.. S. Harrison, Witham Lodge, 374, Lincoln
.. T. Heffernan, Witham Lodge, 374, Lincoln
.. N. Highmore, P.M. Lodge of Benevolence, 459, Sherborne
.. Frederick Heisch, P.S.W. Shakspere Lodge, 116, 324, 388, London
.. J. Hodgson, Belfast, *(3 copies)*
.. J. Hodgkinson, Burlington Lodge, 113, London
.. W. Hodgkinson, Royal Standard Lodge, 730, Kidderminster

Bro. H. F. Holt, P.M. Cadogan Lodge, 188, London
.. T. Horne, J.G. First Lodge of Light, 689, Birmingham
.. —How, Bank of England Lodge, 329, London
.. Jer. Howes, P.M. Lodge of Perseverance, 258, Norwich
.. R. P. Hunt, W.M. Hope and Charity Lodge, 523, Kidderminster
.. W. Imrie, D.C. Bank of England Lodge, 329, London
.. Thomas James, W.M. 786, Walsall
.. C. Johnson, Light of the North Lodge, 69, Londonderry
.. H. Kennet, P.M.and Trea. Social Friendship Lodge, 326, Madras
.. G. E. Lane, I.G. Social Friendship Lodge, 326, Madras
. Capt. G. A. Leach, Royal Engineers, Light of the North Lodge, 69, Londonderry
.. S. H. Lee, W.M. Fitzroy Lodge, 830, London
.. T. Lemale, P.M. Burlington Lodge, 113, London
.. Rev. T. Lindsay, Light of the North Lodge, 69, Londonderry
.. Ignazio Marini, Bank of England Lodge, 329, London
.. F. Crew, Grand Master's Lodge; Sec. Girls' School
.. F. H. M'Causland, Light of the North Lodge, 69, Londonderry
.. M. M'Dowell, P.M. Universal Charity Lodge, 340, Madras
.. Hon. R. Garraway M'Hugh, P.M. Albion Lodge, 762, Castries St. Lucia.
.. J. Melton, S.D. Universal Charity Lodge, 326, Madras
.. Alex. Montague, W.M. 730, Cheltenham
.. T. R. Moore, M.D. Salisbury
.. R. C. Moore, P.J.D. Witham Lodge, 374, Lincoln
.. Thomas B. Morrell, M.A. 730, Oxford
.. T. Morris, S.W. Universal Charity Lodge, 326, Madras
.. J. Motherwell, M.D. Light of the North Lodge, 69, Londonderry
.. E. Mullins, P.M. Bank of England Lodge, 329, London
.. Rev. Erskine Neale, Kirton Rectory, Suffolk
.. J. S. Newton, Stew. First Lodge of Light, 689, Birmingham
. S. Noble, P.M. Pythagorean Lodge, 93, Greenwich
. Patrick O'Leary, P.M. Hope and Charity Lodge, 523, Kidderminster
. Capt. W. O'Neil, Light of the North Lodge, 69, Londonderry
.. S. Ormsby, Light of the North Lodge, 69, & S.D. 640, Londonderry
.. Wm. Palmer, Royal Standard Lodge, 730, Kidderminster
.. P. V. Pereira, Calcutta
.. Osmond G. Phipps, Ramsgate, P. M. 149, and 621, Provincial Grand Organist, Kent
. E. Preston, Light of the North Lodge, 69, Londonderry
. Samuel Pruce, Trea. Hope and Charity Lodge, 523, Kidderminster
. T. Pryer, P.M. Oak Lodge, 225, London
. W. H. Read, P.M. Zetland in the East Lodge, 748, Singapore
. James Rettie, W.M. 54, Aberdeen
.. C. Rice, P.M. 339, Boston
.. J. Y. Robins, S.D. First Lodge of Light, 689, Birmingham
.. T. Rounce, P.M. Lodge of Fidelity, 813, Southwold
.. J. K. Sanford, P.M. St. John's Lodge, 703, Rio de Janeiro
.. J. Sharp, P.M. Shakspere Lodge, 356, Warwick
.. G. T. W. Sibthorpe, S.W. Witham Lodge, 374, Lincoln
.. J. Sims, J.W. 707, Handsworth

Bro. J. Sims, J.D. First Lodge of Light, 689, Birmingham
·· H. S. Skipton, M.D., S. W. Light of the North Lodge, 69, Londonderry
·· E. D. Smith, W.M. Bank of England Lodge, 329, London
·· J. Colson Smith, P.M. Zetland in the East Lodge, 748, Singapore
·· H. S. Smith, Bristol, (4 *copies*)
·· George Southall, P.M. Hope and Charity Lodge, 523, Kidderminster
·· R. Spencer, P.M. Bank of England Lodge, 329, London, (50 *copies*)
·· C. Stroughill, S.D. Bank of England Lodge, 329, London
·· — Tamberlik. Bank of England Lodge, 329, London
·· W. M. Tayler, W.M. First Lodge of Light, 689, Birmingham
·· R. Taylor, P.M. Social Friendship Lodge, 326, Madras
·· J. F. Taylor, Stew. First Lodge of Light, 689, Birmingham
·· W. Taylor, J.D. Hope and Charity Lodge, 523, Kidderminster
·· C. T. Townsend, P.M. Perfect Friendship Lodge, 522, Ipswich
·· R. Toynbee, Witham Lodge, 374, Lincoln
·· W. Trimble, Light of the North Lodge, 69, Londonderry
·· W. G. Turner, J.D. Social Friendship Lodge, 326, Madras
·· T. Vesper, jun. W.M. Yarborough Lodge, 812, London
·· T. Wallace, W.M. Lodge of Fidelity, 813, Southwold
·· J. G. Waller, Bank of England Lodge, 329, London
·· J. Walmsley, P.M. Liverpool
·· F. G. Warrick, S.D. Bank of England Lodge, 329, London
·· Z. Watkins, P.M. Bank of England Lodge, 329, London
· John F. White, P.M. Castle Lodge, 36, London
·· W. Foster White, P.M. St. Paul's Lodge, 229, London
·· J. Whitmore, P.M. Bank of England Lodge, 329, London
·· E. G. Willoughby, P.M. 701, 782, Birkenhead
·· M. C. Wilmot, Tyler. Social Friendship Lodge, 326, Madras
·· M. Woodcock, Witham Lodge, 374, Lincoln
· W. L. Wright. P.M. and Trea. Bank of England Lodge, 329, London

TO THE BRETHREN OF THE FOLLOWING

Subscribing Lodges.

Light of the North, 69, Londonderry
Lodge 196, Ireland
Universal Charity, 340, Madras
Royal Sussex, 589, Belize, Honduras
Royal Victoria, 649, Nassau, Bahamas
St. John's, 703, Rio de Janeiro
Zetland in the East, 748, Singapore
Albion, 762, Castries, St. Lucia
Social Friendship, 326, Madras
Lodge of Instruction, Bristol
Lodge of Instruction, Norwich
Lodge of Instruction, Liverpool

AND TO THE OFFICERS AND BRETHREN OF EVERY LODGE THROUGHOUT THE UNIVERSE.

DEAR BROTHERS AND FRIENDS,

HAVING arrived at the age of nearly threescore years and ten, my labours in behalf of the beloved institution of Freemasonry must be considered as drawing to a close; and I have therefore thought it a duty to open the present volume with an Address to you, as a grateful return for the uniform courtesy which the Craft have testified towards me. These labours have not been actuated by motives of a pecuniary nature, for I have derived very little profit from my masonic publications; and my chief reward has been in the kindness and attention which I have received from the fraternity at large. I have never been troubled with an ambition to accumulate riches; nor have I ever been overburdened with wealth, or greatly inconvenienced by its absence.

> Man wants but little here below,
> Nor wants that little long.

To secure your approbation has been my chief aim, and the possession of it constitutes the utmost limit of my ambition.

The elements of a general address are so diversified, that the canon proposed by Churchill, however it might fail in a formal treatise, will apply excellently well here.

> This I hold,
> A secret worth its weight in gold,
> To those who write as I write now,
> Not to mind where they go, or how,
> Through ditch, through bog, o'er hedge and stile,
> Make it but worth the reader's while;
> And keep a passage fair and plain
> Always to bring him back again.

Some authors construct their prefatory introduction as a programme of the book; some to conciliate the reviewers; and others, more venturesome, hurl at the critics their unmitigated defiance; like the sailor, who having occasion to pass over Bagshot Heath in a chaise, and being told that there were "hawks abroad," deliberately taking a pistol in each hand, he thrust his feet through the front windows crying out, "down with bulk heads, and prepare for action."

My opening address to you, brethren, will be more modest. I have been too long before you, and have received too many of your favours, either to dread a severe sentence, or to feel the necessity of flattering you into good humour. It is well known that while a favourable review of any work passes unnoticed by the multitude, an unfavourable one is sought after with avidity,—circulated amongst the author's personal friends with persevering industry, and frequently perused with the greatest unction,—so much better pleased is poor human nature with hearing abuse and vituperation, than it is with quiet approval; as we often observe a number of people collect together to witness a street quarrel, who will disperse when the dispute ceases, and the conversation assumes a peaceful tone.

For this reason it might be prudent on my part to conciliate criticism; but as my writings contain no severe reflections upon others, I am unwilling to doubt of their candid reception; and after a literary career approaching to half a century, it is too late for me to entertain much apprehension for the fate of a volume which is intended as the completion of a series, and the winding up of a masonic life. Like the Mosaic pavement of the lodge, my pilgrimage has abounded in variegated scenes of good and evil; and success has been chastened and tempered by mortifying reverses. Fast friends I have had many, and bitter enemies not a few; and honours and rewards on the one hand, have been balanced by vexation and trouble, and the basest ingratitude for essential services, on the other.

When I first entered the lists as a masonic writer, it

was intended, in a great measure, for my own private amusement; and the popularity of my earliest publications was entirely unexpected, as they were composed by snatches amidst the pressure of heavy and ceaseless duties, which neither my feelings nor my interest could induce me to neglect. Like Doctor Syntax, I set out in search of the picturesque, and, to my great surprise, found it solid fame. An event, too trifling to be recorded, originated my first publication, and consequently produced all the rest; as Pascal very pleasantly attributed the revolutions which took place in the world during the reign of Cleopatra to the longitude of her nose.

At that early period I had formed a plan in my own mind, which was intended to demonstrate the capabilities of Freemasonry as a literary institution. It was generally understood to be pursuing one unvarying round, circumscribed within a very narrow compass; including the ceremonies of initiation, passing, and raising, with a prescribed lecture for each degree; touching, indeed, upon morality and several liberal sciences, but determinate upon none.

To convince the reading public that Freemasonry possessed within itself references of a more exalted character, and that it actually contained the rudiments of all worldly science and spiritual edification, I contemplated working out, in a specified cycle, a detailed view of its comprehensive system of knowledge, human and divine. The plan was extensive, and the chances were, that it would share the fate of that gigantic edifice on the plains of Shinar, which was intended to scale the heavens, and never be completed. But the mind of youth is elastic Hope urged me on, and enthusiasm lent its powerful aid to encourage me to persevere; and with the blessing of the Great Architect of the Universe, I have now the pleasure to present you with the concluding volume, by which the cycle is perfected and the cope stone laid.

And I may say, in the poetical language of a Most Excellent Master,

All hail to the morning that bids us rejoice;
The temple's completed, exalt high each voice;
The cope stone is finished, our labour is o'er;
The sound of the gavel shall hail us no more.

It may be interesting to exhibit the entire plan in detail.

It will occur to every thinking brother, that such an undertaking, to be perfect, must necessarily embrace History and Antiquities; Rites and Ceremonies; Science and Morals; Types and Symbols; Degrees and Landmarks; and, above all, it would require to be shown what connection the Order bears to our most holy religion; and how far it recommends and enforces the duties which every created being is bound to observe in his progress from this world to another and a better. To all these points my attention has been extended; and for the purpose of exemplifying them by a regulated process, I have, at convenient intervals, issued from the press a graduated series of publications, each advancing one step beyond its predecessor, and, like the progressive terms of a syllogism, contributing their united aid to produce an intelligible conclusion.

The first step was to show the antiquity of the Order, and somewhat of its early history: for this was the only basis on which all subsequent reasoning could be securely founded; and in the absence of this footstone, the entire fabric, like the enchanted palace of Aladdin, would be unsubstantial and endure but for a moment. I therefore published a work on the Early History and Antiquities of Masonry from the Creation to the building of Solomon's temple, as an acknowledged period from which the history of the Order is clear and intelligible; including dissertations on those permanent Landmarks of Masonry, the Creation, the Fall, the Deluge, the calling of Abraham, the vision of Jacob, the deliverance from Egyptian bondage, the construction of the Tabernacle, the passage of the river Jordan, the contest of Jeptha with the Ephraimites, and the construction of the Temple.

This attempt being well received, although I was comparatively a stranger to the fraternity in general, having merely published a few masonic sermons, as the Grand Chaplain for Lincolnshire, previously to this period, which, it is extremely probable, were unknown beyond the limits of the Province, I was encouraged to proceed in my design.

Still the foundation was not complete. It was necessary to show clearly to what religion, if any, the present system of masonry was analogous. On this question I came to the point at once, like Hippothadee in Rabelais, "without circumbilivaginating about and about, and never hitting it in the centre," and unhesitatingly pronounced it to be Christianity, not only from internal evidence, but also from the following considerations.

Freemasonry is unquestionably a cosmopolitical institution, and therefore must have an affinity to a religion which is applicable to all times, and adapted to every people that have at any time existed in the world. These data are true with respect to Christianity and to no other religion that ever existed. The patriarchal dispensation was incipient Christianity. The holy men who lived before the time of Moses were all justified by the same principle of faith in God's revelation. They looked to the same blessings in futurity that are revealed in the Gospel; and it is this principle of faith which will constitute their reward as well as ours at the day of judgment.

Again, the revelation of the Jewish religion was another wide step towards the introduction of Christianity. It was the second degree of perfection. And in this belief the saints and prophets who came after Moses offered themselves freely to all the persecutions of the world in proof of their faith in a Deliverer to come; and hence the holy men under the law are held forth by the Apostles of Christ as examples to their followers. And the whole design of the Epistle to the Hebrews is to show that the faith of the patriarchs, both before and after Moses, was the same as ours, though their worship was of a different form.

For these reasons, as the Christian religion extended over all time, and shall, at the appointed period, universally prevail over the whole earth, it alone can apply to a cosmopolite institution like Freemasonry. The principal events in the Jewish history are types of Christ, or of the Christian dispensation. But these events form permanent and unchangeable landmarks in the masonic lectures. Therefore the lectures of masonry are Christian. This decision is borne out by a manifesto of a foreign Grand Lodge, which contains the following characteristic passage, "Masonry may be made the means of accomplishing the commands of the Great Architect of the Universe. He who is the best Christian, the most faithful man, will be also the best Mason. So let it be in the profane world and in church relations,—live in brotherhood and peace."

It is an artifice of the enemies of masonry, such as Mr. E. C. Pryer, Major Trevilian, and all others of the same school, to insinuate that masonry is anti-christian, that it may become unpopular and lose its influence; but that the cry should be echoed by those who pretend to be acquainted with its genuine principles, surpasses my ingenuity to comprehend. It is the very point to which Weishaupt was desirous of bringing the disciples of Illuminism. He taught that "genuine Christianity is no popular religion, but a system for the elected; that Jesus communicated the higher sense of his doctrine only to his most intimate disciples; that the latter had propagated this system among the primitive Christians by means of the *disciplina arcani;* taught it in the mystic schools of the Gnostics, Manichæans, and the Ophites, in a two-fold manner, viz., exoterically and esoterically; that at the last, after many migrations, and concealed in hieroglyphics, it had become the property of the Order of the Freemasons." Meaning to infer that Christianity was a system of Ophiolatreia, and preserved only in the arcane mysteries of the Freemasons; and that consequently pure Christianity was an unsubstantial vision.

As Bishop Watson said of the opponents of Christianity, I repeat of the enemies of our noble Order. "I have often wondered what could be the reason that men, not destitute of talents, should be desirous of undermining the authority of [Freemasonry], and studious in exposing, with a malignant and illiberal exultation, every little difficulty attending it, to popular animadversion and contempt. I am not willing to attribute this strange propensity to what Plato attributed the Atheism of his time—to profligacy of manners—to affectation of singularity—to gross ignorance, assuming the semblance of deep research and superior sagacity;—I had rather refer it to an impropriety of judgment respecting the manners and mental acquirements of humankind in the first ages of the world."

To place this matter on the proper basis, and to show the opinion of eminent brethren of the last century, I published the STAR IN THE EAST, in which I endeavoured to show the absolute connection between Freemasonry and religion from the testimony of masonic writers; from the fact that the historical portion of the lectures bears a direct reference to Christianity; from the coincidence between the morality of masonry and that of our holy religion; and the symbolical reference of its general mechanism to the same faith.

The rapidity with which the first edition of this little

work was exhausted, and the testimonies I received from intelligent brethren in every part of the United Kingdom, to its value as a standard Text Book of Masonry, convinced me that I had been correct in my opinion of the universal belief that the present system of Freemasonry is analogous to the Christian religion.

I cannot throw odium or even doubt on the cross of Christ; nor can I allow any contempt to be cast on that sacred atonement by which I trust to inherit the kingdom of heaven, either by my silence or connivance. I will admit my Hebrew brother into a mason's lodge—I will exchange with him freely all the courtesies of civil and social life; but as he will not abandon his faith at my command—neither will I. We each pursue our own path, under the consequences of our own free choice, like Thalaba and his companion in the cavern of Haruth and Maruth. It is a false species of liberality which influences the feelings of many good and estimable men at the present day, and induces them to concede, out of respect to the prejudices of others, what they ought to hold most sacred. Ask your Hebrew brother to lay aside his prejudices, and eat with you—and he will reject your proposal with abhorrence. And he acts on a correct and laudable principle—for it is in accordance with the injunctions of his religion.

A writer in Sharpe's Magazine asks, "what is liberality? for this is, after all, the question. We should not perhaps greatly err in representing it as a complex idea, embracing the virtues of courtesy, beneficence, charity in judgment, and self-denial in conduct. St. Paul was the first example of it, after the only perfect example of all good. His speech before Agrippa, his Epistle to Philemon, are instances of a refined courtesy; his beneficence and self-denial are alike instanced in his laborious journeys, and his manual exertions to minister unpaid; his charity and kind judgment are the soul of all his conduct. Yet St. Paul would have gained no credit for liberality in our day; for he would have made no sacrifices to spread Judaism or Gnosticism; and further, he did his best to overturn both, while showing every kindness to the persons of those who professed them. While he commanded to do good to others, he added, specially unto those which are of the household of faith. Nothing could be more illiberal, according to the principle on which the word is received at the present day; for even if doing good unto all men were admitted on that principle, we must now add—specially unto those who are NOT of the household of faith."[1]

I am far from affirming, however, that the analogy of Freemasonry with Christianity is universally conceded by the fraternity. Our ranks contain many individuals, whose opinions are entitled to respect, who reject the hypothesis as an untenable proposition; and are ready to maintain that the glorious Symbol which forms the subject of this volume is alien to the system of Freemasonry. And they assign as a reason for their theory, that as Freemasonry dates its origin at a period far anterior to the revelation of the Christian scheme, its elements cannot legitimately contain any reference to that great plan for the salvation of the human race.

The argument, however, is inconclusive, because it is at variance with fact. Freemasonry, in whatever part of the globe it may at present exist, contains the emblem before us, sanctioned by all Grand Lodges, and rejected by none. And it is interpreted by a process agreeing with our own explanations; embodied in the authorized Lectures, as propounded by the united wisdom of the two great sections of the fraternity assembled in the Lodge of Reconciliation, which was constructed for the sole purpose of placing the Order on its proper basis, by

[1] Sharpe's Mag., vol. vii., p. 48.

revising the Lectures and regulating the ceremonies on the true model of primitive observance.

Freemasonry must be interpreted according to the form in which it is actually presented to the senses, and not by any hypothetical propositions of what it was or might have been at a given period which is too remote for any records to exist that may explain its mechanism or peculiar doctrines, and respecting which our traditions are too imperfect to lead to any certain result. And the present Lectures of the Order actually contain a pointed reference to all the principal types of Christ or the Christian dispensation which are found in the Hebrew Scriptures, from the creation of the world to the actual appearance of the Messiah, when the sceptre had finally departed from Judah.

The Freemasons of 1720, in the earliest system of Lectures known, explained the masonic phrase, T G A O T U, to mean, "Him that was placed on the topmost pinnacle of the temple;" which applies to Jesus, and to him alone, as no other personage on record was ever placed in that inaccessible situation. The revised Lectures of Bro. Dunckerley, used up to the middle of the century, defined the Blazing Star as "representing the Star which led the wise men to Bethlehem, proclaiming to mankind the nativity of the Son of God, and here conducting our spiritual progress to the Great Author of our redemption." The Hutchinsonian Lectures, used twenty years later, explained the three lights or luminaries by "the three great stages of masonry; the knowledge and worship of the God of nature in the purity of Eden—the service under the Mosaic law, when divested of idolatry—and the Christian revelation. But most especially our Lights are typical of the holy Trinity." And in the system of Lectures which prevailed at the latter end of the century, and up to the union in 1813, the five steps of the winding staircase were represented as indicating "the birth, life, death, resurrection, and ascension of our Lord and Saviour Jesus Christ."

The authorized Text Book of the United States of America confirms this view of the design of Freemasonry; and it will be remembered that the Royal Arch is pronounced by the English Grand Lodge as the completion of the Third Degree. The account of this degree commences thus: "This degree is more august, sublime, and important, than all which precede it. It impresses on our minds a belief of the being and existence of the supreme Grand High Priest of our salvation, who is without beginning of days or end of years; and forcibly reminds us of the reverence due to his Holy Name." And that there may be no mistake in the meaning of "the supreme Grand High Priest of our salvation," the degree is opened by a passage from St. Paul's Epistle to the Thessalonians,[2] "Now we command you, brethren, in the name of our Lord Jesus Christ, that ye withdraw yourselves from every brother that walketh disorderly," &c.

Having thus laid the foundation of my proposed edifice on a solid basis, broad and deep—on the antiquity of its pretensions, and its undoubted reference to an universal religion—as I professed to write for the general information of the fraternity, I now found, as honest John Bunyan has it, that "I must not go to sleep, lest I should lose my choice things;" and, therefore, commenced the superstructure with an explanation of the elementary tenets of the Order, as a preliminary step towards a general view of its claims to a favourable consideration, which might spread throughout the length and breadth of the habitable globe.

No science can be mastered without a competent knowledge of the terms and technicalities by which it is distinguished; and Freemasonry, like Chemistry, will

[2] 2 Thes. iii., 6—18.

be very imperfectly understood, unless the tyro be well grounded in the hidden meaning of the types and emblems in which its occult principles are imbedded and concealed. Canons must be studied ere perfection can be attained. Every one must be an apprentice before he can entertain any pretensions of becoming a master. Without this preparation, no one will ever become an adept in the science of Freemasonry.

To supply this desideratum, the volume called SIGNS AND SYMBOLS was next offered to the masonic public; and it appears to have been fully appreciated by the fraternity, as the first edition of a thousand copies was sold off in a few months. In this work, I went fully into detail on all the acknowledged emblems by which the Order is distinguished, and explained them seriatim; and a particular index enumerates upwards of two hundred Symbols which have been noticed in the twelve lectures which complete the volume. This publication opened a very extensive correspondence with brethren in every quarter of the globe where masonry flourishes, and the testimonies in its favour were so numerous and decisive as to constitute an unequivocal encouragement to proceed in my design, which I did not think it prudent to decline; particularly as H. R. H. the Grand Master, after the work had been submitted to his inspection in MS., kindly allowed it to be dedicated to him; and H. R. H. the Duke of York, the Duke of Leinster, Grand Master of Masonry in Ireland, and several of the nobility and Provincial Grand Masters extended their patronage to it.

The plan being now fairly opened, it appeared to me to be necessary, before proceeding further, to obviate an objection which had been raised against the antiquity and originality of the Order, by bringing forward a series of authentic evidences to prove that Freemasonry stands proudly on its own basis, without being indebted to the religious mysteries of heathen nations; for it was confidently affirmed by some respectable authorities, that the death of Osiris, Adonis, or Bacchus, which was celebrated in those institutions, constituted the prototype of the mysteries of Freemasonry. And the identity of one with the other was alleged to be complete, not only from internal evidence, but from the supposed correspondence of facts and ceremonies with the Mithratic celebrations.

Voltaire had treated Osiris, Hiram, and Christ, as fabulous avatars of the same personage. Volney, Professor Robison, and others, had promulgated the same opinions. In answer to all which, Mr. Maurice remarks that, "in the pure and primitive theology, derived from the venerable patriarchs, there were certain grand and mysterious truths, the object of their fixed belief, which all the depravations brought into it by succeeding superstitions, were never able entirely to efface from the human mind. These truths, together with many of the symbols of that pure theology, were propagated and diffused by them in their various peregrinations through the higher Asia, where they have immemorially flourished; affording a most sublime and honourable testimony of such a refined and patriarchal religion having actually existed in the earliest ages of the world;" and this simple mode of faith was Freemasonry in its most primitive form.

Mr. Fellows, an American writer, promulgated the opinion that "the cenotaph, or mock coffin, used in the anniversaries, is typical of the death of the sun in the inferior hemisphere, under the name of Osiris, who is personated under the Hiram of masonry." And De Quincy, an eminent and clever writer of our own country, adds, "in the earlier records of Greece we meet with nothing which bears any resemblance to the masonic institution but the Orphic Eleusinian mysteries. Here, however, the word *mysteries* implied not any occult problem or science sought for, but simple, sensuous and dramatic representations of religious ideas, which could not otherwise be communicated to the people in the existing state of intellectual culture. In the Grecian mysteries, there were degrees of initiation amongst its members," &c.

To combat these erroneous opinions, and to clear the way for future discussion, I published a complete view of the entire system of religious mysteries, as practised in every part of the idolatrous world, under the name of a History of Initiation; which, like the former, passed rapidly through the first edition; and a second was published before the expiration of the year. This work contains a detailed view of the Spurious Freemasonry of India, Egypt, Persia, Greece, Britain, Scandinavia, Mexico, and Peru; thus displaying in one point of view, all the principal mysteries which were practised over every part of the globe, noting their resemblances and peculiarities, to show that they had a common origin, which was dated at a period anterior to the general dispersion on the plains of Shinar, and entirely unconnected with the traditional origin of Freemasonry.

But notwithstanding the most complete demonstration of a case, and however its truth may be apparent to an impartial judge, there are those who are tardy to confess an error in any theory which they have once advocated. It seems on a par with an acknowledgment of mental imbecility, and an incapacity to determine a simple proposition when submitted to their unbiassed judgment; and they are cautious of admitting the least doubt of the soundness of their intellect, lest their literary credit, on which their future success appears to depend, should suffer any diminution; for every man is sensitively alive, and properly so, to the slightest shadow of a stain on his reputation. And hence arises the anomaly of men persisting in error, even after their reason is convinced that they have advocated an unsound hypothesis, although the most honourable course would be to acknowledge it untenable, and candidly regret that a hastly formed opinion should have led them astray. But to return.

A progress thus signalized by unequivocal success was not likely to produce lassitude on my part. I therefore prepared to advance another step in my great design, by an endeavour to elucidate the true philosophy of the Order, to show it *as it is*, and not as it ought to be, according to the expressed opinions of some theorists; to describe its construction, to display its use and tendency, and to enquire whether it has any correspondence with practical religion, and the duties which are enjoined by the Most High, as the test of faith and purity of heart; for science, said the sage Iracagem,[3] "may polish the manners, but virtue and religion only can animate with exalted notions, and dignify the mind of immortality; to neglect the first, is to turn our head from the light of day; but to despise the last, is to grasp the earth when heaven is open to receive us. A wise and prudent spirit will so use the one as to improve the other, and make his science the handmaid of his virtue."

To enunciate the above particulars I now anxiously address myself. Serious doubts had been entertained by the unlearned in its mysteries, whether Freemasonry possessed any rational claims to the character of a Literary and Scientific Institution; and these surmises were strengthened by the consideration that no proofs of it were to be discovered in any authorized publication which was accessible to the general enquirer. The absence of these proofs being elevated into a cogent and unanswerable argument of their non-existence, the

[3] Tales of the Genii, viii.

cowan triumphed in the imaginary abasement of a science (so called) below the level of the most common mechanical art; because they all, itself excepted, could produce abundant evidences of their utility, either in theoretical lectures, or the exquisite perfection of their manual productions; while nothing, as it was urged, appeared on the surface to recommend Freemasonry to public notice, but the external existence of the lodge room, decorated with symbols of a technical and speculative character, which, like the complicated diagrams of a necromancer, might bear an interpretation either puerile or fearful, no one knew which; with an occasional procession to assist at some operative ceremony, which terminated in a banquet; and there irregularities were sometimes exhibited, altogether incapable of extenuation or defence. While public lectures on Freemasonry were unknown, it was deduced that its utility and moral or scientific tendency were questionable, if not altogether imaginary.

Under such circumstances it became necessary to the well-being of the Order, that some attempt should be made to neutralize the effects, if it were found impracticable to defeat the existence of such unfounded assertions. And this could only be accomplished by placing within the reach of every enquirer, who would take the trouble to investigate the truth, some authentic treatise on the peculiar philosophy of the Masonic Order; and my previous publications having established for me an humble claim to the character of an authorized teacher of Freemasonry, it was suggested by several scientific brethren, for whose opinions I entertained considerable respect, that the fraternity looked up to me for some general undertaking which might silence the absurd cavils of our opponents, and place Freemasonry on the broad basis of an acknowledged literary institution.

For this purpose I published Twelve Lectures on the

THEOCRATIC PHILOSOPHY OF FREEMASONRY, in which I entered minutely into an examination of the speculative character of the institution as a system of Light and Charity; and of its operative division as an exclusively scientific pursuit which had been practised from the earliest times in every country of the world. This enquiry was followed up by an historical account of the origin, progress, and design of the Spurious Freemasonry. I then took a view of the origin and use of hieroglyphics; and not only exemplified the symbols used in those spurious institutions which had attained the most permanent celebrity in the ancient world, but endeavoured to show that the true Freemasonry in all ages was "veiled in allegory and illustrated by symbols." The union of speculative and operative masonry then became a subject of discussion; and I concluded with a detailed disquisition on the form, situation, ground. extent, and covering of a lodge, as well as an exemplification on the beauties of Freemasonry, in which I attempted to show that its peculiar ceremonies and observances had been judiciously selected, rationally maintained, and highly advantageous to those who are versed in their moral and symbolical references.

These disquisitions became extremely popular, and the edition was speedily exhausted. I was next called on to show what masonry was actually doing at the present period, for the purpose of evincing that it was not exclusively theoretical, but that whatever had been advanced in theory was verified in practice. The fourteenth edition of PRESTON'S ILLUSTRATIONS, which I had edited in 1829, had been sold off, and the fraternity were anxious to see the history of masonry which it contains brought down to the present time; for the Order had flourished for the last ten years beyond all former precedent, and it was thought expedient to place its transactions on permanent record. The HISTORY OF FREEMASONRY from 1829 to 1840 was accordingly prepared, and I have reason to believe that its publication was attended with beneficial consequences to the Craft in general, as it was received with marks of favour by the universal concurrence of the fraternity.

During the whole of this period I had been a constant and regular correspondent to the Freemasons' Quarterly Review; and my articles in that useful miscellany were applied to the general dissemination of masonic knowledge, in its Speculative, Operative, and Spurious divisions; diversified by an occasional essay on its charities and its amusements. And I communicated the more readily and cheerfully with this journal, because I think that the great modifications which have manifested themselves in the opinions of men towards Freemasonry of late years, would not have been produced without the efficient aid of this powerful engine. It has effected a wonderful revolution of opinion in favour of the Order by mild and gentle reasoning; and has rendered extensive benefit to the Craft by diffusing information on the rules of discipline and practice, as well as on the public transactions of the lodges. To the blemishes of the system it has applied the actual cautery with singular effect. The patient has, indeed, winced under the operation; but the cure is in progress, and the treatment, however unpalatable, is wholesome and salutary, and cannot fail to be productive of the most beneficial results.

I had now arrived at a period when, however unmerited on my part, my literary fame stood high with the masonic public, and I could produce letters from every quarter of the globe in testimony of the utility and general estimation of my labours, of which I am, indeed, proud, although they have failed to make me vainglorious. By nature humble and unassuming, it is a difficult task to draw me out for the purpose of lionizing. The attempt has been made at sundry times and seasons, but with very little effect. I still remain snugly ensconsed in my "hollow tree," and have no taste for the distinction of exhibiting before popular assemblies. Perhaps I may be afraid of the fate of Don Quixote, who, at the highest pinnacle of his glory, when he had converted two flocks of sheep into rival armies, and saw them so clearly as actually to describe the armour of the knights and the devices on their shields, and at the moment when he contemplated a triumphant charge, was almost knocked on the head by an ignoble brickbat. However this may be, my attempts have been unceasing to restore a sublime Order to its legitimate place in public estimation, from which, as I well recollect, it had somewhat retrograded at the period when the two great divisions were united in 1813, probably from the petty jealousies, and continued disputes of the brethren belonging to each of these sections, which nullified its claim to the characteristic of brotherly love.

My next undertaking was a great work on the Historical Landmarks of the Order. The design was comprehensive, and was intended to embrace particular explanations, both historical, scientific, moral, and ceremonial, of Symbolical, Royal Arch, Ineffable, and Sublime Masonry, including the military orders and degrees. But in my own case, as I had been before the masonic public as an author nearly thirty years, and enjoyed the good fortune of being received with distinguished favour, it might be rationally imagined that I had employed my time unprofitably if I was not qualified for the undertaking.

Besides, as the Eidolon confessed to Capt. Clutterbuck, in the Introductory Epistle to the Fortunes of Nigel, "while I please the public, I shall probably continue it merely for the pleasure of playing; for I have felt as strongly as most folks that love of composition which is perhaps the strongest of all instincts, driving the author to the pen, the painter to the palette, often without either the chance of fame or the prospect of reward." And I was not without hope that my project would be well received, if it were executed creditably.

The labour required for the collection of materials for such a design, which was originally intended to be comprised in fifty-two lectures, with an abundance of explanatory notes, would doubtless be very great; but part of the work had already been accomplished during the researches which had been previously made for former publications. My stores were far from being exhausted; and my previous training in masonic lore had not been unproductive; yet, I speak it in sorrow, when the work was fairly launched, and the preliminary number before the public, in which I had committed myself by a solemn pledge to carry it forward to the end, I experienced feelings somewhat similar to those of Pope when he had undertaken to produce a translation of Homer. "What can you expect," he says to his friend Jervas, "from a man who has not talked these five days? Who is withdrawing his thoughts, as far as he can, from all the present world, its customs, and its manners, to be fully possessed and absorbed in the past? When people talk of going to church, I think of sacrifices and libations; when I see the parson, I address him as Chryses, priest of Apollo; and instead of the Lord's Prayer, I begin,

"God of the silver bow," &c.

While you in the world are concerned about the Protestant succession, I consider only how Menelaus may recover Helen, and the Trojan war be put to a speedy conclusion. I never enquire if the queen be well or not, but heartily wish to be at Hector's funeral. The only things I regard in this life are, whether my friends are well; whether my translation goes well on," &c.

In like manner my thoughts, wishes, and aspirations were all on masonry, and nothing but masonry. It formed the subject of my labours by day and my dreams by night, during the two long years it was in hand, from the Introduction to the Index; occupying a space of nearly fourteen hundred pages. But the patronage of the Earl of Zetland, the M. W. Grand Master, by giving authority to the work, was a full and adequate recompence for all my anxiety and toil. These volumes embrace a full and copious exposition of the doctrine and discipline, ceremonies and symbols, not only of blue masonry, but every order which had been at any time assimilated with it, whether justly or unjustly, to the amount of nearly a thousand degrees, including the blue, red, and black masonry of our own country, and the Ancien et Accepte of the Continent and the United States, besides the speculative systems of the Orders of Bruce, the Temple of Clermont, of Strict Observance, of Mount Tabor, of Zinnendorff, Swedenborg, Tschoudy Mesmer, Cagliostro, and many other empirics who invented systems for their own personal emolument; together with Adoptive, Swedish, Adonhiramite masonry, and other varieties in different countries which were identified with the Order.

About this time certain imputations were cast upon Freemasonry, which, by their mild and insidious nature, and constant repetition, were calculated, as the eternal droppings of water will in time penetrate and wear away the most solid substances, to do more serious injury to the cause than all the absurd charges that were ever urged respecting the addictions of its members to the ridiculous fancies of Rosicrucianism and diablerie; because they attacked it on the side of its religion and morality, and would have converted us into a swarming hive of infidels. These charges originated in India, and had been deliberately concocted, and circulated in the public journals of the country; putting on, like Iago, the form of civil and humane seeming, for the better compassing of their hidden loose affections; with the design of withdrawing all good and pious men from the ranks of masonry, and preventing such persons, who, it will be believed, constitute our best and most valuable members, from seeking admission amongst us. And this effect it succeeded, in a few instances, to produce.

The objections, four in number, were stated as follows:—1. That a true Christian cannot, or ought not, to join in masonry, because masons offer prayers to God without the mediation of a Redeemer. 2. That masonry inculcates the principles of brotherly love and charity to those peculiarly who have been initiated into the Order; whereas such acts, to be acceptable to God, should proceed from a love of him reconciled to mankind through the sacrifice of Christ; any other motive being not only not acceptable, but sinful. 3. That the mention of the Lord's name in the lodge is a contravention of the third commandment. And 4. That the Protestant Church of England knows nothing of the society of Freemasons, and therefore it is a desecration to suffer any section of that society to appear in the character of masons within the walls of its sacred edifices.

My attention was called to the subject by a zealous mason in India, who stated all the charges seriatim in a letter to Dr. Crucefix, with a request that they might be forwarded to me for refutation. Accordingly I discussed them fully in a series of papers in the Freemasons' Quarterly Review, as they were certain to reach their destination through the medium of that periodical. They were afterwards transferred to a pamphlet called AN APOLOGY FOR THE FREE AND ACCEPTED MASONS, with additions, in reply to a statement which the Rev. Mr. Blunt, of Helston, in Cornwall, imputed to the Bishop of Exeter, to the effect that "the Church of England knows nothing of the distinctive principles of the society of Freemasons;" which, indeed, may be true in the abstract, but by no means available as a reason for refusing the use of a Christian church for a masonic sermon, because a numerous host of the clergy, with the then Archbishop of Canterbury at their head, belong to the masonic body, and consequently may be supposed to know something of the distinctive principles of the Order.

During the course of my researches for the illustration of the Historical Landmarks, I accidentally met with documents which singularly enough threw considerable light on other points in the history and details of the masonic system, that had hitherto remained in obscurity, and respecting which my enquiries had been previously unsuccessful. Several intelligent brethren had frequently expressed an earnest desire to be satisfied on certain undetermined questions which I was anxious to resolve. The result of these discoveries was given to the world in two pamphlets on the ORIGIN OF THE ROYAL ARCH, and on the unfortunate SCHISM which divided the Craft into two independent sections for more than half a century; both of which I have reason to believe were satisfactory, and will set all speculation on each of these subjects at rest for ever.

I also published a series of letters on the JOHANNITE MASONRY, addressed to the Earl of Aboyne, P. G. M. for Northamptonshire and Huntingdonshire, on which two hostile opinions exist amongst the fraternity, although masonic lodges were always dedicated "to God and holy St. John," by our ancient brethren; and Scottish masonry acknowledges that holy Apostle as its peculiar patron and tutelary saint. Besides, our annual festivals are enjoined to be celebrated on the day of St. John the Baptist or St. John the Evangelist; and down to a very recent period these two holy men were universally considered the great parallels of the Order.

To place this important question on its proper basis,

and to afford materials for an impartial discussion of its merits, I first endeavoured to refute a few objections which had been urged against the masonic parallelism of the two St. Johns; then I instituted an enquiry whether the patronage of masonry was originally vested in the two St. Johns during the last century or at any earlier period; and gave my reasons for considering each of these saints separately as a patron of masonry. And after a copious explanation of the parallelism, I enquired whether the patronage of masonry in the hands of these two Christian saints be strictly conformable with the construction and character of the Order; and ended with a recapitulation of the whole argument, and a reply to some recent charges which had been published respecting the Rosicrucian origin of the Order.

These charges are of very ancient date, for anti-masons have existed in every age of the world, although they have been recently adduced as novelties, and there is nothing new under the sun. Passing over Sanballat and his associates, the first anti-mason we read of in Christian times was called Simon Magus, who mistook, as all his followers have done to the present day, the system of Christianity for a species of Rosicrucianism, by the exercise of which the Apostles were enabled to perform miracles and alter the ordinary course of nature. His fate is well known. He was followed by Barjesus, struck with blindness by St. Paul; the Nicolaitans, and the Gnostics. Then came Hymenæus, Marcion the teetotaller, Alexander the coppersmith, and the actors in the ten Roman persecutions. A goodly company; with whose proceedings and character the modern cowan appears ambitious to be classed.

One of his most famous prototypes is the celebrated Manes, who, like Simon Magus and the anti-masons of our own times, endeavoured to identify the system of Light with the occult philosophy and the practice of judicial astronomy, which was afterwards called Rosicrucianism. He, like his predecessor Marcion, recommended total abstinence from intoxicating liquors, and substituted in their stead various amulets and charms as a protection from danger. Another worthy of the same class was the impostor Basilides, whose Powers and Intelligences, good and evil angels, with his Serpent Serapis, Abraxas, and three hundred and sixty-five demons, our opponents would fain identify with Freemasonry; but the utter absurdity of his doctrines and practices constitutes an undeniable proof that they have no alliance with its principles.

The catalogue might have been extended to the present time, terminating with the worthies Barruel and Robison, Soane and E. C. Pryer; for every age abounds with them; including Voltaire, Paine, and Carlisle in the old world, and Morgan, Allyn, Stone, and Bernard in the new. In company with such worthy associates, par nobile fratrum, the cowan will doubtless consider it honourable to persevere; and it may therefore be expected that the Order will never be without opponents, to restrict its means of doing good.

During the latter part of my masonic career, I have received frequent and particular enquiries respecting masonic ceremonies of public and private occurrences, about which the information has been scantily imparted, and consequently an exact uniformity is scarcely to be found. On public occasions particularly, such as processions, footstones, &c., a great diversity of practice has existed in different localities; and visiting brethren have found it difficult to reconcile the anomalies which they have discovered in various lodges, where accident or design has induced them to be present. Enquiries into the practice of antiquity respecting ceremonies on which the Book of Constitutions is silent, and consequently much is left to the knowledge or discretion of masters of lodges, have been numerous and pressing, and descended to the minutest particulars; even to the form and colour of every article of the dress of a Master Mason, from the hat on his head to the buckle of his shoe; the former being supposed to be necessarily triangular, and the latter an oblong square.

Although I have never omitted to comply with such requests, but have always placed myself at the command of the fraternity, as a reasonable tribute of gratitude for the continual marks of uninterrupted favour with which I have been honoured; yet it was at length suggested that if all points of enquiry were collected, and categorically arranged in the form of a Hand-Book, it would constitute an acceptable present to the Craft, as an useful Manual which might be at every brother's disposal, and referred to on all occasions as an authority from which there could be no appeal. And accordingly I took the hint, and issued the Book of the Lodge, which I have no doubt will be esteemed a necessary companion to every brother who is desirous of obtaining, at little expence and trouble, correct information on the rites and ceremonies of the Order.

It may not be amiss, in this gossiping Address, which is confined to no particular subject, if I subjoin a few observations on the symbolical and actual habiliments of a Master Mason, as enjoined by authority in other times; which I think I have not enlarged on elsewhere. At the revival in 1717, it was directed—and that there might be no mistake about the matter, the canon was inserted by Anderson and Desaguliers in the earliest code of lectures known, that the symbolical clothing of a Master Mason was, "skull cap and jacket yellow, and nether garments blue." After the middle of the century he was said to be "clothed in the old colours," viz., purple, crimson, and blue; and the reason assigned for it was, "because they are royal, and such as the ancient kings and princes used to wear; and we are informed by sacred history, that the veil of the Temple was composed of those colours;" and therefore they were considered peculiarly appropriate to a professor of "a royal art." The actual dress of a Master Mason was, however, a full suit of black, with white neckcloth, apron, gloves, and stockings; the buckles being of silver, and the jewel suspended from a white ribbon by way of collar. This disposition prevailed until the Union in 1813, when it was ordered that in future the Grand Officers should be distinguished by purple, the Grand Stewards by crimson, and the Master Mason by blue, thus reverting to "the old colours" of our ancient brethren.

It will have been observed, that throughout these desultory remarks, no notice has been taken of those subordinate parts of an author's employment, which consist in editing and illustrating the works of other men, although the undertaking is of a more laborious nature than writing an original work. It requires deep consideration to dive into the private thoughts of others, and penetrate the hidden meaning of abstruse passages which apply to another state of society. The masonic writings of our brethren of the last century are few in number, and had become scarce and inaccessible; although they are of great value, inasmuch as they delineate the gradual improvements of the Order, and mark the process by which it imperceptibly disarmed its adversaries, and converted them into firm and active friends.

For many years after the great revival, Freemasonry was considered a paradox beyond the comprehension of ordinary capacities. As the mystical institution silently forced itself into notice, the world wondered, and some daring spirits ventured to assail it with the shafts of ridicule. Indeed, so much importance was attached to its proceedings, that even Hogarth and Swift did not disdain to join in the hostile array. The clamour was, however, allayed by the judicious efforts of Anderson, Desaguliers, Martin Clare, Calcott, Dunckerley, Smith, Hutchinson, Preston Inwood, and other gifted brothers, who quietly explained its principles, and directed public notice to the virtues which it inculcated, and to the symbols in which they were imbedded and preserved. Many valuable fragments are unfortunately lost, but the Remains are amply sufficient to excite the attention of the fraternity. Under these circumstances, I conceived that an acceptable service would be rendered the Craft, by collecting the scattered rays of Light and bringing them into one focus, that they might contribute their aid to the general illustration of the science.

The above authors left behind them detached pieces on the subject of Freemasonry which are of great value; and they have been collected and reproduced under the general title of the GOLDEN REMAINS OF THE EARLY MASONIC WRITERS; which consist of five volumes on Masonic Institutes, Principles and Practices, Persecutions, Doctrines, and Morality; each volume being introduced by an original Essay on one of the following subjects, viz., the Masonic Literature of the eighteenth century; the social Position of Symbolical Masonry at the same period; Usages and Customs; Masonic Tests; and Cypher Writing. In the mean time, new editions of the entire works of Hutchinson, Preston, and Ashe, were published under my editorial superintendence. In a catalogue raisonné of my masonic labours, these trifles need only be mentioned as forming a series of adjuncts to the general design, like the statues or pictures that adorn the walls of a lodge-room, which is perfect in its construction without them, but more ornamental and pleasing to the eye by their assistance.

The following work completes the series, and constitutes the cope-stone of the edifice, by exhibiting a view of the ultimate resting-place to which all men aspire, and which offers itself prominently to the eye of the mason every time he enters the lodge. The steps which lead to it are gradual and progressive.

> By just degrees they every moment rise,
> Fill the wide earth, and mount unto the skies.
>
> Pope.

The Holy Bible forms the basis of this great moral machine. It rests on the altar of Omnipotence, and proclaims the rewards of faith and practice; while the Ladder connects earth with heaven, where the perfect mason hopes to consummate his worldly labours, and receive the recompence of his fidelity.

I now feel like the architect, who, seeing that his plan in the erection of a magnificent edifice is nearly completed, entertains some fears lest the finishing ornaments should deform the whole building. My Lodge has been erected according to the established rules of art; the floor has been consecrated, the internal decorations disposed in order, and not a single indispensable ceremony has been omitted, which might tend to confer the attribute of perfection on the whole design; but as the hawk, when certain of his quarry, sometimes suffers the fate which he tries to inflict, I must take especial care that I do not impale myself on the heron's threatening bill. The covering is the most important portion of a lodge, and, to make it perfect, requires a judicious combination of skill and judgment. In this volume the experiment is made, but it needs the decision of the fraternity to determine whether it will be attended with success.

That a fund of useful information is spread over the volume, which is not contained in any of my former works, will admit of neither doubt nor denial. And be the judgment of the brethren what it may, I shall not imitate the example of the Archbishop of Granada, after his fit of apoplexy, who expostulated with his critic, by observing, "Say no more, my child," said he, "you are yet too raw to make proper distinctions. Know that I never composed a better homily than that which you disapprove; for my genius, thank heaven, hath as yet lost nothing of its vigour. Henceforth I will make a better choice of a confidante. Adieu, Mr. Gil Blas, I wish you all manner of prosperity, with a little more taste."

For my own part, I am open to fair and gentlemanly criticism; and although I may be mortified at finding my hard-earned fame melt away like an icicle in the sun, yet I shall not complain if you, my dear brethren, pronounce it to be your deliberate opinion that my late severe indisposition has impaired my faculties, and disqualified me for a masonic writer. It is rather late in life to divest myself of habits of thinking and acting which I have fostered for nearly half a century, and which have constituted almost the only source of pleasure and gratification in which I have freely indulged during that extended period; but I shall endeavour to lay them aside in cheerful acquiescence with the decision of those who are better judges than myself, if the opinion should prove to be unfavourable. I entertain, however, a sanguine hope that you will consider the *covering* to be at least equal to the rest of the fabric, and that the cope-stone adds beauty rather than deformity to the work. Should my anticipations be correct, your approval will be a cheering reflection at the latter end of a life spent in the service of the fraternity.

The above confessions may throw some light on the origin and design of publications which have long been familiar to you. The most satisfactory method of dis playing the usefulness of a science, is not by merely showing the extent of its application, but also the diversity of subjects which it embraces; and this has been my object throughout the whole of my publications. If a pursuit is to be estimated according to its results, Freemasonry may be safely classed amongst the most comprehensive of human sciences, and therefore the best adapted to the state of man on earth. And it is idle to object that its fruits do not appear in every initiated brother. For if it were worth while to investigate the number of dabblers in any given science, we should find that those who really excel bear no greater proportion to those who fail, than may be traced in Freemasonry. The sciences of music, drawing, architecture, chemistry, and various others, have myriads of tyros, but few masters; and the *æs Dodonæum*, the loudest talker in these wordy days, is frequently the most shallow.

It may be readily admitted, that there are a great number of masons who are contented with very trifling acquirements in the art. So much the worse for them. But it will not follow that we possess no shining examples of excellence, although, from the nature of the institution, they are necessarily confined to the atmosphere of their own particular localities; for no lodge can flourish for any length of time except it possesses some intelligent master spirit to conduct its proceedings. Freemasonry is not a science that admits of itinerant lecturing; and therefore it cannot be expected that the uninitiated public know much about it; and consequently their conjectures are very wide of the truth. They shoot their arrows wildly, and seldom hit the mark. Guessing is an unsatisfactory employment, and they are more profitably engaged in the macaronic diction of the Grubbian Expostulantiuncula,

> Qui pro niperkin clamant, quaternque liquoris
> Quem vocitant homines Brandy, superi Cherrybrandy,
> Sæpe illi long-cut, vel short-cut (returns) flare tobacco
> Sunt soliti pipos.

Freemasonry is a secret institution; and its peculiar benefits are limited to its own body. And although we act upon the ancient principle of *procul hinc quivis scelestus*, yet our lodges are open to all good and worthy men, and our mysteries are hidden from none but those whose presence would be of doubtful benefit, either to themselves or the Order. We do not invite adherents, and therefore none can be disappointed. But we rejoice when men of name in science or literature solicit admission amongst us, because we may confidently anticipate that the expectations of every person who possesses taste and judgment will be fully realized, and the pursuits of masonry be congenial to his mind.

I keep lingering over my task, and continue to scribble for lack of moral courage to pronounce the word "FAREWELL" to those dear brethren and kind friends by whom I have ever been so well received and bounteously treated. *Jucundi acti labores.* And still the benediction must be uttered; for "the best of friends must part," and the most intimate and beloved connections will be severed in the end. In my various publications I have endeavoured to redeem the Order from the charge of frivolity, which was brought against it in the last century, by showing its applicability to many of the sciences—I have portrayed its literary character—I have pointed out the various sources of amusement and instruction of which it is the author and dispenser; and in this, my closing work, I have shown how, in concurrence with other causes, its sincere professors, through the merits of the Great Architect of the Universe, may find their way to another and a better world. My labours cannot have a more satisfactory termination. I am verging on that period which our Grand Master David pronounces to be "labour and sorrow," soon to pass away and be gone; and it is extremely probable that the fraternity will hear little more about me, except in an occasional Paper in the Freemasons' Quarterly Magazine and Review!

I now subscribe myself,
Dear Brethren and Friends,
Your obliged and faithful servant,
GEO. OLIVER, D.D.

SCOPWICK VICARAGE,
Oct. 1*st*, 1850.

THE SYMBOL OF GLORY.

THE

SYMBOL OF GLORY.

LECTURE I.

Epistle Dedicatory

TO

BRO.	EDMUND A. RAYMOND, ESQ.,	G. M.
—	REV. GEO. M. RANDALL,	D. G. M.
—	JOHN J. KORING, ESQ.,	S. G. W.
—	THOMAS M. VINSON, ESQ.,	J. G. W.
—	CHARLES W. MOORE, ESQ.,	G. SEC.
—	THOMAS TOLMAN, ESQ.,	G. TREA.

Of the G. L. of Massachusetts, U. S.

Dear Brethren, Friends and Associates,

As the first and chief Grand Lodge in the United States of America, it will be needless for me to assure you of the high value which I place on the masonic dignity that you have conferred upon me in a manner not merely flattering to my feelings, but peculiarly honourable as an unequivocal testimony of your appreciation of my masonic labours.

It is an exalted step, to which my humble ambition had never, even in thought, aspired; and I am proud to have this public opportunity of testifying my gratitude.

I would convince the gainsayers that masons entertain a strong sense of obligation for favours received; and show them that in the Lodge, as well as in the world, the incitements to a career of virtue do not fail to bring forth the fruits of good living, to the honour and glory of T G A O T U.

In my intercourse with mankind on the subject of Freemasonry, I have been accustomed to class its opponents under three distinct heads. 1. Those who hate masonry because it is a secret institution, without being able to assign an adequate reason for their dislike. 2. Those who live in the neighbourhood of an ill-conducted lodge, and see the evil consequences which result from carelessness on the one hand, or intemperance on the other. And 3. Those who are desirous of admission, and do not possess the requisite courage to encounter the presumed terrors of initiation.

These classes are equally destitute of the most essential virtues of the masonic order, faith, and hope, and charity. Believing nothing—hoping nothing—like the magician, Happuck, in the fairy tale, they entertain the most inveterate feelings towards Freemasonry, because it favours the cause of virtue; and against which their objections are unsupported by the slightest shadow of evidence. All argument with them is therefore useless. One of them being asked why he continued to oppose Freemasonry, when, if he would take the trouble to read the publications of the Order, his prejudices would be effectually removed, very coolly replied: "Perhaps so—but I never *do* read!" This puts me in mind of an anecdote of Don Pedro's private confessor, who, when exhorting the Portuguese to battle, assured them that if they should fall, they would, that very night, eat their suppers with the blessed. With this assurance they went to battle and were defeated, the holy confessor being the first to run away. One of his companions shouted to him—"How is this Father? Did you not tell us that those who fell should sup in Paradise?" "Yes," said the confessor, "but I never eat suppers!"

None of the above mentioned classes have any just grounds of complaints; and their tirades against the Order are therefore gratuitous in their motive, and unjust in their end. The divine science is perfectly unobtrusive, it is not forced on their notice; it pursues the even tenor of its way, and interferes with no other society or class of men whatever. Where, then, lies the grievance? How are they injured? Does it monopolize any of their privileges—does it deprive them of any advantage—does it supersede any of their enjoyments?

Nothing like it. It offers no disturbance to their habits of thought; it prevents no course of study, proscribes none of their amusements, nor defeats any of their plans, whether domestic, civil, or religious. Where, then, does the shoe pinch? This question is answered by the story of the banishment of Aristides from Athens, because his sense of honour and justice was too great to allow him to prostitute his principles at the bidding of a successful rival.

But, perhaps, they complain that if masonry, as is asserted, possesses any peculiar benefits and advantages, they ought to share in them. It is a fair presumption; but it contains a full refutation of their own arguments and objections. For the benefits of masonry are open to their acceptance. They are refused to none who are worthy; and it will scarcely be contended that they ought to be conferred alike on the good and the bad. It would be like casting our pearls before swine; as they might thus be converted to an evil purpose, and reflect equal disgrace on the institution and themselves.

If all the professors of our noble and sublime science would endeavour to merit the character of good and worthy masons, by a regular attendance on the duties of the Lodge; by studying the peculiar principles of masonry, which I have embodied in the present Volume; and by practising in their several stations the precepts which are there inculcated, then would our opponents see and acknowledge the pre-eminent beauties of the Order, and be fully convinced that Speculative Masonry is something more than an empty name.

In order to effect this purpose, I have taken the liberty,

M. W. Grand Master,
And my worthy peers,
The Officers of the Grand Lodge,

To dedicate to you the following Lecture, containing

some suggestions which, it is hoped, will merit your attention; and to subscribe myself,

With great respect,
And fraternal affection,
Your obedient Servant and Brother,
GEO. OLIVER, D.D.,
Past D. G. M. of the Grand Lodge of Massachusetts

SCOPWICK VICARAGE,
June 1, 1849.

Lecture the First.

On the present state of the Masonic Science.

"Yn that tyme, throggh good Gemetry,
Thys onest craft of good Masonry
Wes ordeynt and made yn thys manere,
Ycownterfetyd of thys clerkys y fere;
At these lordys prayers they cownterfetyd Gemetry,
And gaf hyt the name of Masonry—
Far the most oneste craft of alle."

ANCIENT MASONIC MS.

"Laws convenient, proper, and effective at the time in which they were made, have not been altered to accord with the altered circumstances of Freemasonry, and the extension of the Lodges and localities of the fraternity. Such alterations must, however, be made in Freemasonry in accordance with the landmarks of the Order, which in this as in all other cases must be kept holy and inviolate."

FREEMASONS' QUARTERLY REVIEW, 1847.

IN my letters to the Earl of Aboyne, P. G. M. for the counties of Northampton and Huntingdon, on the Johannite Masonry, I threw out a hint, that, on account of the altered state of society since our present Lodge Lectures were framed by the Lodge of Reconciliation, and enjoined by authority in 1814, a new revision was become necessary, to meet the requirements of an improved mode of thought arising out of the many extraordinary and unexpected sources of information which have been thrown open to the fraternity, by the rapid strides that science is making at the present period, and the many new vehicles for the propagation of knowledge which have become accessible by means of literary and philosophical societies, reading rooms, mechanics' institutes, and the exertions of itinerant lecturers to familiarize the most abstruse scientific and philosophical subjects to the capacities of all classes of mankind, which unite their aid to enlighten the understanding, and improve the morals of the present generation.

Since the publication of these Letters, I have given my undivided attention to that particular subject, and am now fully convinced that such a revision would be attended with essential benefits to the Order. The masonic experience which I acquired during my occupation of the chair of a private Lodge for eleven years in the whole, succeeded by the sole management of a large and populous Province for nearly the same length of time, enables me to speak with some degree of confidence, on all subjects connected with the details, as well as the general principles of the Order. And having observed, with feelings of sorrow and regret, its sensible decline in my own Province since the period of my decadence from that high office, a few remarks on the above subjects may neither be unacceptable nor inappropriate.

Some years ago, the Grand Lodge of Ireland issued a paper of Queries to every private Lodge under its jurisdiction, that the general opinion of the Craft might be collected "as to the best means of improving the Order of Freemasonry." Amongst these queries we find the following. "Is the Order improving or declining? If declining, to what cause do you attribute its decay? What is the prevailing opinion among persons not of the Order respecting masonry? Is masonry reputable or disreputable in your neighbourhood? What measures would you recommend for improving the state of the Order?"

If some such course were adopted by other Grand Lodges, it might lead to a very useful result; for we frequently hear the enquiry repeated by the non masonic world, that, in the present stirring times while science has been so rapidly on the advance, what has Freemasonry accomplished? This is a question which every right minded brother would rejoice, for the credit of the Order, to see triumphantly answered by a detail of the advantages which mankind have derived from its successful exertions, or the happy application of its principles to the general benefit of society.

Now it is well known that the operation of Freemasonry is confined, in a great measure, to morals; although it is not without a just claim to some degree of merit as a teacher of science. And if we trace its progress for the last thirty years in every quarter of the globe where it flourishes, we shall find it fairly entitled to its share in the polite literature of the day, sanctioned by Grand Lodges, and patronized by wise and benevolent Grand Masters. These writings have contributed not a little to the general amelioration of the morals, and improvement in the tastes and manners of men which distinguish the nineteenth century of Christianity.

The system of Freemasonry at the present day, is marked by an adherence to the good old custom, so strongly recommended and assiduously practised by the masonic worthies of the last century, and imitated by many other public bodies of men, of assembling the brethren of a Province annually under their own Banner, and marching in solemn procession to the House of God, to offer up their thanksgivings in the public congregation for the blessings of the preceding years; to pray for mercies in prospect, and to hear from the pulpit a disquisition on the moral and religious purposes of the Order. It is to this custom that we are indebted for those invaluable treasures of masonic literature that are exhibited in the printed discourses of our clerical brethren. As for instance, those of our Reverend brothers Harris and Town, (U. S. of America); Inwood and Jones, (Kent); Haverfield, (Hampshire); Dr. Carwithen, (Devon); Dr. Orme (Lincolnshire); Grylls, (Cornwall); Erskine Neale, Freeman, (Suffolk); Walker, (West Yorkshire); Percy, (Dorset); Roberts, (Monmouth); Gilmour Robinson, (West Lancashire); Buckeridge, (Staffordshire); Broderip, (Somersetshire); Taylor, (Cheshire); Archdeacon Mant, (Ireland); Eyre Poole, (Bahamas); Hovenden and Ruspini, (Bengal); and many other talented and pious brethren whose names it would be tedious to enumerate.

This custom is sufficient of itself to ensure the popularity of the Order, and create a respect for its holy principles in the public mind. I much regret that a practice so consonant with the original design of masonry should have been discontinued in my own Province, and exchanged for other public observances, which, though they may be innocent, are a novel introduction; and in my opinion, an application of divine masonry to purposes that were never contemplated at its original institution. This hint may not be without its use in other localities; and if the one must needs be done let not the other be omitted.

The above custom would also be a means of promoting and encouraging that great attribute of the order—EQUALITY. But lest this principle should be confounded with the communism and fraternization which have worked such irreparable mischief in other countries, it may be useful shortly to explain its design and reference as used by the Free and Accepted mason. The system of equality observed in a mason's lodge, teaches the doctrine of mutual wants and mutual assistance, and destroys the unsocial vice of Pride, by the operation of which one man is induced to despise his brother, as though he was not formed of the same clay as himself, although he may be greatly his superior both in talent, virtue, and usefulness. Freemasonry is essentially democratic in its construction, and strikes at the root of this pernicious vice, which wrought the destruction of Nimrod and Nebuchadnezzar, Bali of Hindoostan, and Shedad of the Paradise of India, by laying it down as an axiom that "we are all equal by our creation, but much more so by the strength of our obligation;" and that "we meet on the level and part on the square."

Now, according to the doctrines of the Order the level demonstrates that we are descended from the same stock, partake of the same nature and share the same hope; and that though distinctions among men are necessary to preserve subordination, yet no eminence of station can make us forget that we are brethren, and that he who is placed on the lowest spoke of fortune's wheel, may be entitled to our regard; because a time will come, and the wisest know not how soon, when all distinctions, except that of goodness shall cease; and death, the grand leveller of human greatness, reduce us to the same state.

The lodge lectures are copious in carrying out this

principle, that there may exist no possibility of misunderstanding it. They instruct us that in the lodge a king is reminded, though a crown may adorn his head and a sceptre his hand, the blood in his veins is derived from our common parent, and is no better than that of his meanest subject. The statesman, the senator, and the artist, are there taught that equally with others, they are exposed by nature to infirmity and disease; that unforeseen misfortunes may impair their faculties and reduce them to a level with the meanest of their species. This checks pride, and incites courtesy of behaviour. Men of inferior talents, or who are not placed by fortune in such exalted stations, are also instructed in the lodge to regard their superiors with peculiar esteem, when they discover them voluntarily divested of the trappings of external grandeur, and condescending, in the badge of innocence and bond of friendship, to trace wisdom and to follow virtue, assisted by those who are of a rank beneath them. Virtue is true nobility, and Wisdom is the channel by which virtue is directed and conveyed; Wisdom and Virtue only, mark distinction among masons.

Nothing can more vigorously contribute to the banishment of pride from a mason's lodge, than such disquisitions. But to prevent the benignant principle of Equality from being prostituted to unworthy purposes, and used as a vehicle for any improper assumption of character, the ancient Charges provide that in the lodge the brethren are to pay due reverence to the Masters, Wardens, and Fellows; and out of the lodge they are directed to salute one another in a courteous manner, calling each other brother, freely giving mutual instruction as may be thought expedient, without being overseen or overheard, and without encroaching upon each other, or derogating from that respect which is due to any brother, were he not a mason; for though all masons are, as brethren, upon the same level, yet masonry takes no honour from a man that he had before; nay, rather it adds to his honour, especially if he had deserved well of the brotherhood, who must give honour to whom it is due.

As a vice, nothing is more intolerable, or more debasing than pride; by which I mean that exclusive feeling which elevates one member of society, in his own opinion, to an imaginary distinction above another of the same rank, and perhaps superior endowments. For this reason it is formally repudiated in the system of Freemasonry. Our Grand Master, King Solomon, was more urgent in his condemnation of this vice than on any other subject. He declares his hatred of "pride and arrogancy, and a froward mouth;"[1] and for this reason, because it produces contention,[2] brings a man to shame,[3] and certain destruction.[4] Indeed, throughout the whole of the Sacred Scriptures, this vice is unequivocally prohibited as the bitter parent of all evil. Pride was not made for man. Our blessed Saviour classes it with adultery, fornication, murder, theft, covetousness, deceit, blasphemy and foolishness.[5] And St. Paul adds that, "he who is lifted up with pride falls into the condemnation of the devil.[6]

In a word, of all the evils which have been introduced by the wicked spirit as the curse of man in his civil and social state, pride is the most pernicious. Every single vice is bad, but pride is the consummation of them all. And hence Freemasonry, that benevolent, and truly amiable science, has most unceremoniously banished it fróm the lodge, and sung its requiem; for it is a moral leprosy, by which the soul is spotted and defiled, and filled with "wounds, and bruises, and putrifying sores." Even the heathen, who were ignorant of the benignant principles of true religion, believed its existence to be hostile to the peace and comfort of society. Tacitus says, Multos qui conflictari adversis videantur, beatos; ac plerosque, quanquam magnas per opes, miserrimos; si illi gravem fortunam constanter tolerent, hi prosperâ inconsultè utantur. And the ethnic poet, Horace, promulgated the same doctrine, when he said,

Non possidentem multa vocaveris
Rectò beatum. Rectius occupat
Nomen beati, qui deorum
Muneribus sapienter uti,
Duramque callet pauperiem pati.

Dr. Doune illustrates this vice by these judicious reflections, which are worth preserving. "Death comes

[1] Prov. viii. 13. [2] Ib. xiii. 10. [3] Ib. xi. 2.
[4] Ib. xvi. 18. [5] Mark vii. 21, 22. [6] 1 Tim. iii. 6.

equally to us all, and makes all equal when it comes. The ashes of an oak in a chimney are no epitaph of that oak, to tell me how high, or how large, that was; it tells me not what flocks it sheltered while it stood, nor what men it hurt when it fell. The dust of great persons' graves is speechless, too; it says nothing, it distinguishes nothing. As soon as the dust of a wretch whom thou wouldest not, as of a prince whom thou couldest not look upon, will trouble thine eyes if the wind blow it thither; and when a whirlwind hath blown the dust of the churchyard into the church, and the man sweeps out the dust of the church into the churchyard, who will undertake to sift those dusts again and to pronounce—this is the patrician, this is the noble flower;—and this is the yeoman, this is the plebeian bran."

I have been rather diffuse upon this unmanly vice, because it is so positively prohibited in a mason's lodge: and I think also that if the teaching of Freemasonry on this particular point, were carried out in practice amongst mankind, it would strengthen the bond of union which cements man to his fellow, and thus become of the most essential service to society in general.

But the most distinguishing glory of Freemasonry is Charity; which, indeed, constitutes the peculiar characteristic of the age in which we live. Public institutions for benevolent purposes have sprung up in every metropolis and provincial town throughout the world, and there is no class of destitution which is now unprovided with a retreat where their sorrows are assuaged, and their wants supplied. The good Samaritan is every where at work. In this point of view also Freemasonry must be regarded as the agent of unbounded good. To its male and female orphan schools, and fund of Benevolence, which have long been in active and beneficial operation, we have added not only an Asylum for the worthy aged and decayed members of the fraternity, and an Annuity Fund for the benefit of the same class of destitute persons; but a projected establishment for the permanent support of the widows of indigent Freemasons has been mooted in Grand Lodge, with the best wishes of the Craft for its happy termination, and I do not entertain the slightest doubt but it will ultimately be accomplished. In addition to all these noble institutions, we have private masonic funds for benevolent purposes in many of the lodges both of the old and new world.

These details will clearly evince the claims which masonry has on the community at large; and that the active part she has sustained in forwarding the benevolent enterprizes by which the present age is distinguished, merits public approbation. We appear to be on the eve of some great and organic changes; whether for good or evil, the Great Architect of the Universe can only determine. But it behoves Freemasonry to take such steps in the great drama of life, as to secure, if it be possible, the predominance of good. She ought to occupy the foremost rank in the work of amelioration, to watch over the best interests of the public, and endeavour to prevent the inconsiderate and unwary from being misled by the false glitter of unsound theories on the one hand, and hollow professions on the other, which are sure to terminate in disappointment and disgrace, and perhaps in consequences of a much more serious nature.

If Freemasonry do not thus exert the influence she undoubtedly possesses for the benefit of humanity, her social claims will be nullified, and her pretensions pronounced to be an empty boast. It is quite clear, from a consideration of the uniform and gradual alterations and improvements in the details of Speculative Freemasonry by successive grand lodges, that it was never intended to be stationary. The science had no prescribed lectures before the revival in 1717, but every Master of a Lodge exhorted his brethren to the practice of moral virtue, in short and extemporaneous addresses, according to his own capacity, and adapted to the comprehension of the brethren and the state of the lodge. An old masonic manuscript of the tenth century, as is supposed, which may be found in the Old Royal Library in the British Museum, contains ample directions for this purpose. It strongly recommends the brethren to offer up their prayers regularly to God through Christ; to do their duty to each other, and to be constant in their attendance on the divine services of the church. It concludes by advising,

Play thou not but with thy peres,
Ny tell thou not al that thou heres,

Dyskever thou not thyn owne dede,
For no merye, ny for no mede;
With fayr speche thou myght have thy wylle,
With hyt thou myght thy selven spylle.

* * * * * * *

Cryst then of hys hye grace,
Geve yow bothe wytte and space,
Wel thys boke to conne and rede,
Heven to have for yowre mede!
Amen! Amen! So mot hyt be,
Say we so alle per charyte.

In the Lansdowne MS. in the British Museum, (Burleigh Papers, N. 98, Art. 48,) we have another specimen of this moral teaching which is of great antiquity. The Master is there directed "in the name of the Father, Son, and Holy Ghost, to be true to God and holy church, and to use no error or heresy; to be a true liege man to the king, and to do to every brother as he would like to be done to himself. That he shall keep truly all the council of the Lodge or of the Chamber; be no thief; true to the Master; and call his fellows by no other name than brother. That he shall not injure or pollute his brother's wife or daughter; and shall honestly pay for every thing he has."[7]

The earliest authorized Lectures which I have met with, were compiled from such ancient documents as these, and arranged in a catechetical form by Desaguliers and Anderson, as early as 1720. And this form was adopted because it was considered to be more useful in assisting the memory, and affording an efficient remedy against forgetfulness or want of attention, than any other plan. The questions and answers are short and comprehensive, and contain a brief digest of the general principles of the Craft, as it was understood at that period. The First Lecture extended to the greatest length, but the replies were circumscribed within a very narrow compass. The Second was shorter, and the Third, called "The Master's Part," contained only seven questions, besides the explanations and examinations.

If, under such an imperfect system, Freemasonry had

[7] The same Paper contains many other charges for the regulation of conduct, most of which, however, may be found in the 15th Ed. of Preston, p. 71, and see F. Q. R. 1848, p. 142.

not been susceptible of improvement, it could not have stood its ground, during the rapid progress of a taste for refined literature, and the accomplishments of civilized life which distinguished the beginning and middle of the eighteenth century. Intelligent brethren, however, soon became aware that something more than the repetition of a few set phrases and routine explanations, how interesting and important soever they might be in themselves, was required to cement the prosperity, and perpetuate the existence of a great society, which professed to convey superior advantages, and laid claim to a higher character, than any of the numerous antagonistic clubs and coteries of similar pretensions by which it was surrounded. A new arrangement was therefore pronounced necessary in the year 1732, and Martin Clare, A. M., a celebrated mason, who ultimately attained the rank of D. G. M., was commissioned to prepare a course of Lectures, adapted to the existing state of the Order, without infringing on the ancient Landmarks; and he executed his task so much to the satisfaction of the Grand Lodge, that his Lectures were ordered to be used by all the brethren within the limits of its jurisdiction. In accordance with this command, we find the officers of the Grand Lodge setting an example in the Provinces; and in the Minutes of a Lodge at Lincoln, in 1734, of which Sir Cecil Wray, the D. G. M., was the master, there are a series of entries through successive lodge nights, to the following effect; that two or more Sections (as the case might be) of Martin Clare's Lectures were read; when the Master gave an elegant Charge; went through an examination; and the lodge was closed with songs and decent merriment." An evident proof of the authority of Martin Clare's Lectures, or the D. G. M. would not have been so careful to enforce their use amongst the brethren over whom he presided in private lodge.

These lectures were nothing more than the amplification of the system propounded by Anderson and Desaguliers, enlightened by the addition of a few moral references and admonitions extracted from the Old and New Testaments. They also contained a simple allusion to the senses, and the theological ladder with staves or rounds innumerable.

Freemasonry was now making a rapid progress in the island, both in dignity and usefulness; and its popularity was extended in a proportionate degree. Scientific and learned men were enrolled in its ranks, and Martin Clare's Lectures were obliged, in their turn, to give way before the increasing intelligence of the Order. They were revised and remodelled by Bro. Dunckerley, P. G. M., and G. Superintendent for almost half the entire kingdom, whose opinion was considered by the Grand Lodge as decisive on all matters connected with the Craft. In these lectures Dunckerley introduced many types of Christ, and endued the ladder with three principal steps as an approach to the supernal regions, which he called Faith, Hope, and Charity. His disquisition was founded on 1 Cor. xiii.; and he might have had in view the true Christian doctrine of three states of the soul. First in its tabernacle, the body, as an illustration of FAITH; then, after death in Hades, Sheol, or Paradise, as the fruits of HOPE; and lastly, when reunited to the body in glory, about the Throne of God, as the sacred seat of universal CHARITY. The original hint at a circle and parallel lines, as important symbols of the Order, has been ascribed to him.

Thus the Lectures remained until towards the latter end of the century, when Hutchinson in the north, and Preston in the south of England, burst on the masonic world like two brilliant suns, each enlightening his own hemisphere, and each engaged in the meritorious design of improving the existing Lectures, without being conscious that his worthy cotemporary was pursuing the same track. There are reasons for believing that they subsequently coalesced, and produced a joint Lecture, which, though regarded at first with some degree of jealousy, as an unauthorized compilation, was at length adopted, and carried into operation by the concurrent usage of the whole fraternity. This course of Lectures was in practice till the reunion in 1813, and I believe there are still many Lodges who prefer them to the Hemming or Union Lectures, and still continue their use.

With all these facts before us, it is clear that Freemasonry has undergone many changes since its revival after the death of Sir Christopher Wren. The essentials remain the same, but the details have sustained considerable modifications, and are susceptible of still greater improvement. He who ascends the Masonic Ladder, must not tarry at the Portal of Hope, if he wishes to attain the summit. If we are anxious to practise ourselves, or to disseminate for the benefit of others, the poetry and phylosophy of masonry, it will be necessary to show that such progressive alterations may be safely made, without any violation of the real ancient landmarks, or incurring the risk of weakening its hold on the purest affections.

The opinion of our late Grand Master, H. R. H. the Duke of Sussex, was favourable to the views here exhibited. He publicly declared in Grand Lodge, that consistently with the laws of masonry, "so long as the Master of any Lodge observed the Landmarks of the Craft, he was at liberty to give the Lectures in the language best suited to the character of the lodge over which he presided."[8] And as an illustration of his opinion, the Lodge of Reconciliation was authorized to revise and reconstruct the Lectures which were in existence at that period. Under these circumstances, if some slight alterations and improvements were made in the working details of the Order at the present day, with the sanction of the Grand Lodge, I should anticipate the happiest results from the measure.

But the question will be asked, how is this to be accomplished? By what process is such a desirable object to be attained without an invasion of Landmarks, which are so strictly guarded by a fundamental Bye-law, that their integrity cannot be violated without inflicting some serious injury on the institution? The process is simple, and I think practicable; and even if it be attended with some trifling disadvantages, they would be amply compensated by improvements which might be effected under a judicious modification of the lectures.

Thus if the Landmarks, and such portions of the Lectures of each degree as are indispensable to the purity and character of the Order, were drawn out carefully and judiciously in the shape of a series of moral axioms, and divided into degrees, sections, and clauses, constructed

[8] Quarterly Communication, Dec. 1819.

with an equal regard to brevity and perspicuity, and accompanied by a strict injunction that *every brother shall be perfectly acquainted with each before he be admitted to a superior degree*, it appears highly probable that the most beneficial results would be produced. It may, indeed, be imagined, that under such a regimen many brethren would not advance beyond the first degree. I am of a different opinion. The test might discourage indolent and careless candidates; but it would invite and augment the initiations of men of higher character. The facilities afforded by our present *qualifications*, fill our ranks with brotherhood who do us little credit; and the society would be really benefited by their absence. A lodge consisting of a dozen scientific members, would be more respectable, more useful, and more popular, than if it were filled with an uncounted number of sots, or even with dull prosaic brothers who are indifferent to the poetry and philosophy of the Order.

I should certainly anticipate no diminution of numbers under such a course of strict and wholesome discipline. The only perceptible effect would be, to improve the character of the brethren, by creating a spirit of enquiry and discrimination, which would tend to make it their sole aim, as masons, to increase their knowledge, purify their minds, and prepare themselves, by the morality of science, for greater perfection in another and a better state of existence. In our lodges, some brethren are always unfortunately to be found, with whom refreshment is the great attraction and the primary stimulus to their attendance at our stated meetings; but on the improved principle which I would recommend, refreshment, although by no means to be dispensed with, would constitute a secondary motive, while it contributed to give a zest to the theoretical discussions and practical enjoyment which result from the social intercourse of congenial minds.

The only difficulty which appears to attend the above plan, would be in the construction and arrangement of a digest that should meet the rquirements of every section of the Craft; because in a matter of such importance, the concurrence of every Grand Lodge in the universe should be obtained, that a perfect uniformity in work might prevail.

Every institution, to be perfect, should be consistent

with itself. And hence the insufficiency of the present lectures may be questioned. It is therefore desirable that the attention of the fraternity should be fairly awakened to the subject, that they may take the premises into their most serious consideration, and endeavour to place Freemasonry on so substantial a basis, as to constitute the unmixed pride of its friends and defenders; and defy the malice of its traducers and foes, if any such are still to be found amongst those who are indifferent to its progress.

It appears to me that all difficulty would vanish, and a satisfactory arrangement of the various matters at issue might be obtained, if the Grand Lodge were to appoint a Committee composed of brethren resident in London, augmented by delegates appointed from the Provinces, to enquire into all the varieties in the different systems of lecturing throughout the masonic world, and report upon them *seriatim*. And with respect to the Landmarks—as very few points of difference were included in the original system, it would remain an open question whether, by an attempt to reconcile every variety of subsequent introduction, the real Landmarks of the Order would be at all invaded. I shall decline pronouncing any positive opinion on this point, but leave it entirely to the judgment of others.

But should the adoption of any such measure be deemed expedient, the Grand Lodge would not be expected to pledge itself to the absolute sanction of an incipient Report of the Committee, which could scarcely be free from errors. It would be competent to receive the Report; but I should doubt, in a matter of such vital importance, whether that section of it which usually meets in Freemasons' Hall, consisting chiefly of the Masters and Wardens of the Metropolitan Lodges, would be willing to decide the question without a formal appeal to such members of the Grand Lodge as reside in the country, comprising a great majority of its body.

At this stage of the proceedings the Report would be naturally transmitted to the G. M. of each Province, for the consideration of local committees consisting of the Masters and Wardens of the Lodges, with the P. G. M. at their head, and any other scientific brethren out of office, whom they might think proper to associate with them. The Reports from each of these minor bodies, being transmitted to the Grand Lodge, should be subjected to a new committee for collation and revision, and embodied in a general statement of the entire results. A Draft of this being forwarded to all the Provincial committees for their approval, should be finally submitted to the Grand Lodge, who would then, after other preliminaries had been arranged, be in a condition to pass a decisive Resolution on the subject. Communications should be forwarded to the Grand Lodges of Scotland, Ireland, America, the Continent of Europe, and all other places where they exist, accompanied by a detail of the steps which had been taken for the purification of the Order; recommending the alterations to their notice, and soliciting their concurrence. And as there appears to be an universal desire throughout the whole masonic world for some uniform system of working, an opposition to the measure is scarcely to be contemplated. Effectually to prevent such a result, however, it might be advisable to communicate with the foreign Grand Lodges during the progress of the proceedings, soliciting their fraternal suggestions; and a Draft of the final Resolution ought also to be submitted to each of them for approval, before it passed into a law which should be for ever binding on the whole fraternity in every part of the globe, under the jurisdiction of the Grand Lodge of England, as it would be the concurrent production of the united wisdom and research of all classes interested in the triumphant progress of the Order.

Under some well organized plan of this nature, I am sanguine enough to entertain a certain anticipation of such results as would be generally satisfactory; and enable Freemasonry to produce a visible and genial effect on the taste, literature, and morals of the age.

A regular and authentic Text Book being thus provided to preserve the uniformity of the Order throughout the universe, every Master of a Lodge should be directed, either by himself or some other well informed brother of his appointment, to select a passage from this genuine fountain of truth, and deliver an original Lecture each Lodge night for the edification of the brethren; after which a *viva voce* examination should take place; or, which would in some instances be better, a general conversation on the subject which had been thus selected. Such temperate discussions would excite interest and attention; and the energies of individual brethren being thus brought out, much useful information would be elicited; and a permanent impression would be made on the minds of the Junior brethren, which would tend to cement a love of the institution; produce a regular attendance of the members; and be every way advantageous to society at large.

The times in which we live are peculiarly characterized by a deep research into the causes of things, and bold speculations for the improvement of science; and while electricity and chymistry, steam and gas, and machinery of every kind, are earnestly engaged in a contention for superiority, Freemasonry must not pause upon the threshold;—while the world moves on in an uninterrupted course of improvements, Freemasonry must not stand still; for if she hesitates ever so little—time will pass, and she will be distanced in the race.

I have thrown together these few preliminary observations, for the purpose of showing that a taste for the poetry of Freemasonry is necessary, to enable even an initiated brother to extract the honey from the comb, and to imbibe the sweets which the system so abundantly furnishes. If such a feeling were universal amongst the Craft;—nay, if a few talented brethren even, in every private lodge, were in a position to devote a small portion of their time to its cultivation, the most beneficial results would soon be displayed, in the increasing influence of the Order, and its popularity amongst all ranks and descriptions of men.

LECTURE II.

Epistle Dedicatory

TO

BRO.	E. G. PAPELL, ESQ., J. G. W. &	W. M.
—	THOMAS MORRIS,	S. W.
—	CHARLES F. BROWNE,	J. W.
—	WILLIAM CLARKE,	P. M.
—	HENRY KENNET,	TREA. & P. M.
—	WM. BOYD,	SEC.
—	JOHN MELTON,	S. D.
—	WM. GEO. TURNER,	J. D.
—	JAS. G. LAWRENCE,	STEWARDS,
—	J. ARNOLD HICKEN,	

Of the Lodge Social Friendship, No. 326,
Fort George, Madras.

W. Sir and Dear Brethren,

I embrace this public opportunity of assuring you how highly I am gratified by the distinction you have conferred upon me in electing me an honorary member of your Lodge with the rank of a Past Master, because it is an unequivocal testimony that you appreciate at some little value the services I have humbly endeavoured to render to the greatest of all human institutions; although

I am afraid it is more in accordance with your kindness and partiality, than the intrinsic value of the publications to which you have, in such flattering language referred.

I am, indeed, enthusiastically attached to an Order which, in my humble opinion, has been the means of conferring many essential benefits on mankind; not only by the munificence of its members, and the extensive usefulness of its numerous charities, but by the infusion into general society of that refined morality which is taught in the lodge, and, like the genial rays of the Sun in nourishing the productions of nature, has contributed, in no slight degree, to that high toned principle, and correct mode of thinking and acting which distinguish the fortunate times in which we live.

But Freemasonry has a still higher boast, which not only constitutes the pride of its members, but also claims the serious consideration of those who have not had the advantage of initiation into its mysteries. *It forms a step on the road to heaven.* For, in addition to the means and opportunities of acquiring a knowledge of the faith and practice of our holy religion, which the Free and Accepted mason possesses in common with the uninitiated, he has also the advantage of masonic instruction, which the latter do not possess. In the lodge, virtue is arrayed in her brightest form; the practice of Christian morality is strongly recommended and enforced; and the attentive mason is taught, by a series of interesting disquisitions, that if he devotes himself to the observance of the Cardinal Virtues, and is guided by the sacred principles of Honour and Mercy;—if he ascends the staves or rounds of the theological Ladder, by the practice of Faith, Hope, and Charity, he will attain to a residence in the mansions which have been prepared for him by the Most High, to whom be glory for ever and ever.

It is on such considerations as these that my attachment to Freemasonry has been founded. I have adhered to its principles and proclaimed its excellence, amidst evil report and good report, for a long series of years; and I trust that the opinion I have formed of its moral superiority is substantially correct, and will remain unimpaired till T G A O T U shall, in his own good time, translate me to another and a better world.

With fraternal greetings and remembrances, I beg leave respectfully to offer the following Lecture on the Poetry and Philosophy of Masonry,

And to subscribe myself,
Worshipful Sir,
And respected Brethren,
Your obliged,
And humble Servant,
GEO. OLIVER, D. D.,
Hon. Member of the Lodge Social Friendship, Madras.

SCOPWICK VICARAGE,
July 1, 1849.

Lecture the Second.

On the Poetry and Philosophy of Freemasonry.

"Oh, Love fraternal! principle divine!
One touch of thee makes erring nature shine
With the pure radiance of angelic grace
That ting'd with glory Adam's undimm'd face;
Bids strife depart to reign with fools and slaves,
Whose creeds are narrow as their joys and graves!
By thy bless'd power behold one common band
More wonders working than a fairy's wand.
Columbia, Albion, Caledonia, Gaul,
Erin, and Cambria bid their banners fall;
All lands wherein thy influence is felt
Into one universal nation melt."

FROM THE ADDRESS AT THE 12TH ANNIVERSARY FESTIVAL IN AID OF THE ASYLUM FOR AGED FREEMASONS.

It is an universal complaint, and tends to the deterioration of Freemasonry in public opinion, that amongst the numerous initiations which take place annually, so few should be prolific in bringing forth the genuine fruits of the Order. The world view the naked fact with astonishment, and judge unfavourably of the institution from the dearth of eminent characters by which it is distinguished and ennobled. There are not wanting amongst the candidates for admission, men of great talent and high standing in society, and it is very naturally asked, how it happens that their position in masonry so seldom adds to the laurels that adorn their brows?

The question is easy of solution. It is because they have other objects of pursuit which more urgently demand their attention;—or that they do not feel sufficient interest in the subject to enable them to follow up the necessary investigations which may make them perfect in the art;—or that they are not thrown into a masonic society of sufficient calibre to keep their interest alive. In a word, it is because (no matter how it may have arisen) they are not fully imbued with the poetry and philosophy of the Order, but prefer the dull prosaic workings of common life, or entertain mistaken views of its nature and design.

Those extremely talented and useful writers, the Brothers Chambers, speaking on the subject of poetry, say, "poetry may be defined to be the truth inspired by feeling, and breathed into forms of beauty or sublimity. This definition seems to express the essential characteristics of poetry, in all its manifestations; whether the inspired thought be developed in painting, in sculpture, in architecture, (Freemasonry), in music, in language, or in action; they all range themselves under the same formula; for they are but various modes of expressing the same divine principle." And again: "to be a poet, a man must not rest contented with conventionalities and outward shows; with mere arbitrary distinctions of right and wrong, however specious they may appear. He must have that directness and clearness of vision which can at once discriminate between the essential and the accidental; between that which exists in the very nature of things, and that which is merely of artificial growth. An intellectual discrimination, however, is not all that is required. A man may be very acute in detecting fallacies, and even in discerning truth, and yet have but a small claim to the character of a poet. To be a poet, he must not only see beneath the surface of things, but he must feel as deeply as he sees; he must not only see that a thing is true, but he must also feel that it is true; else whatever it may be in itself, or to others, it can be no poetry to him. Let a man possess these two requisites, and if he is but true to himself, if he will but give scope to his own nature, and not fritter away his life and talents by striving to cramp them into some artificial mould prescribed by custom, he will be a poet in the truest sense; if he does not write poetry, he yet cannot fail in that which is often better, for his life will be a real poem, doubtless sadly chequered in its course, but ever eloquent in its significance; ever earnestly striving after the real and innumerable."[1]

[1] Journal, vol. v. N.S. p. 210.

It is for want of being thus deeply versed in the poetry of Freemasonry, that so many, even of the fraternity themselves, differ in their estimate of it. But they draw their opinions from their own private feelings and propensities rather than from any inherent property of the Order. While the bon vivant considers it to be a society established for the purpose of social convivialities, and the man of the world throws it aside as frivolous and useless, the more studious differ in opinion whether it be Christian or Jewish, moral or religious, astronomical or astrological. And all this confusion arises from a confined view of its nature and properties, which limits them to one particular point or phasis of the Order, while, in fact, Freemasonry is cosmopolitical, and embraces the whole region of poetry and philosophy, science and morals. Prejudice, in all its fantastic shapes, is arrayed against us; which, as is well observed by Mrs. S. Hall, in one of her useful moral tales, is the more dangerous, because it has the unfortunate ability of accommodating itself to all the possible varieties of the human mind. Like the spider, it makes everywhere a home. Some of our glorious old fellows—South, or Taylor, or Fuller, or Bishop Hall—has it somewhere, that let the mind be as naked as the walls of an empty and forsaken tenement, gloomy as a dungeon, or ornamented with richest abilities of thinking; let it be hot, cold, dark, or light, lonely or inhabited—still prejudice, if undisturbed, will fill it with cobwebs, and live, like the spider, where there seemed nothing to live upon.

While these shades of difference agitate the members of the society, we are no longer surprised that the uninitiated should wander so much out of their way to satisfy their curiosity as to the real design of the Order. What is masonry? This is the great and important question which has puzzled the heads of all the uninitiated from the day of its first establishment to our own most curious times.

What is masonry? I could give fifty definitions of it if I choose to be communicative; but I should consider myself "courteous overmuch" were I to furnish the cowan with too great a portion of information at once. He would be gorged into a plethoric habit of mind, which would set him a cackling like a young pullet after she has laid her first egg, and hops round the farm yard in an ecstacy of joy to tell her companions what a feat she has done. I shall give him only this one definition at present, and he may muse and meditate upon it at his leisure. Freemasonry is a triangle upon a triangle, placed in the centre towards the rising of the sun; chequered with the opus grecanicum, circumscribed with scroll work, permeating through the Sephiroth, and graduating to a perfect heptad.

There! Let the cowan digest that, and I will then impart some further instruction to edify his mind. He may think these are terms of diablerie and ghost raising. But I assure him they are not. It is true, an ancient objection against the Order was that the Freemasons, in their lodges, "raise the devil in a circle, and when they have done with him, they lay him again with a noise or a hush, as they please." Others diverted themselves with the story of an old woman between the rounds of a Ladder; or with the cook's red-hot iron or Salamander for making the indelible character on the new made mason, in order to give him the faculty of taciturnity.[2] I once initiated a Welch Rector, who was full of the Horatian urbanity as he could hold. Alas, he is gone to the world of spirits, and a better man does not occupy his place. He told me before he was *made*, in his off-hand way, that being desirous of a private interview with his Satanic majesty, he sought initiation as the most probable method of attaining his point; for he understood that he was generally found in propria persona at our meetings, and amused the brethren by beating a tattoo on the board with his hoofs!!! Many a laugh have we had together after his admission, when he knew what the true tendency of masonry was, and the real causes of any extraordinary sounds which might be easily misconstrued.

These, then, constitute some of the absurd conjectures of those unquiet spirits who are ever restless in their search after facts which constantly elude their grasp; and they are as far from enlightenment on the abstruse principles of the Order as were their forefathers, the cowans of the eighteenth century, whose pretended reve-

[2] Anderson, Const. Ed. 1738, p. 227.

lations were fated, each in its turn to disbelief and rejection from all right-minded men. One half the time and talent which they bestow upon the acquisition of illegal knowledge, where their toil cannot fail to be fruitless, would, if they had received initiation, like my friend the rector, and their enquiries had been directed into a legitimate channel, have converted them into good and worthy brothers, and given them an insight into the poetry and philosophy of masonry. This would have secured a permanent satisfaction to their own mind, and conferred upon them the approbation of the fraternity.

Blanchard Powers, an aged transatlantic brother, in his Prize Essay on masonry, thus describes the benefits which it confers on society. "So sublime and heavenly is the royal art, that it solves all difficulties. It kindles a flame of love in the breasts of those who are at the greatest distance from each other, in consequence of their political and religious tenets. It moderates and subdues the spirit of the fulminating priest; his heart is melted into tender affection towards a brother mason; he presents him the friendly hand, and cordially receives him into his bosom, and addresses him by the endearing appellation of a Brother. Masonry lays men under the most solemn obligation to support the government by which they are protected, and never to encourage disloyalty or rebellion. A mason will risk his life for his brother in the hour of danger, though he may be his enemy in the midst of battle."

An intelligible view of the poetry and philosophy of Freemasonry may be gathered from the lodge Lectures themselves; which describe it as "a peculiar system of morality, veiled in allegory and illustrated by symbols." In the old Lectures this description is explained in every section. The floor of the Lodge symbolically teaches that as the steps of man tread in the devious and uncertain paths of life, and his days are chequered by prosperity and adversity, so is his passage through this short and precarious stage of existence. Sometimes his journey is enlightened by success; at others it is obstructed by a multitude of evils. For this reason the floor of the lodge is covered with Mosaic work, to remind us of the precariousness of our situation here; to-day prosperity may crown our labours; to-morrow we may totter on the uneven paths of weakness, temptation, and adversity. Then while such emblems are continually before our eyes, we are morally taught to boast of nothing, but to walk uprightly and with humility before T G A O T U; considering that there is no station on which pride can be stably founded. All men have birth, but some are born to more elevated stations of life than others; yet, when in the grave, all are on the level, death destroying all distinctions. As Free and Accepted Masons then, we ought ever to act according to the dictates of reason and religion, by cultivating harmony, maintaining charity, and living in unity and brotherly love.

In an Icelandic poem quoted by Mallet, we find the following curious picture of the chequered scenes of human life; which, though written at an unknown distance of time, and for the use of a barbarous people, bears a striking resemblance to the peculiar doctrine of Freemasonry on the same subject.

Thà eymdir strida, &c.
When grief oppresses the mournful mind,
And misery's scourges the pale cheeks furrow,
And back the world on thee wends unkind,
And wanton joyaunce derides thy sorrow;
Think, all is round, and will turn anew,
Who laughs to-day may to-morrow rue;
All's equalized.

Again, the illustration of the I M J contains a direction to the same effect. "As the tressel board is for the master to draw his designs on, the better to enable the younger brothers and the more expert Fellow Crafts to carry on the intended building with order, regularity, and success; so may the Holy Bible be justly deemed the tressel board of the Grand Architect of the Universe; because in that holy book he hath laid down such divine plans, and moral designs, that were we conversant therein and adherent thereto, it would bring us to a building not made with hands, eternal in the heavens. The Rough Ashlar is a stone rough as when taken from the quarry, and by the skill and ingenuity of the workmen being modelled and brought into due form, represents the mind of man in its infancy, uncultivated and irregular like this stone, but by the kind care and instruction of parents, guardians, and teachers, in endowing it with a liberal education, the man becomes moralized, and rendered an useful member of society. The Perfect Ashlar is a stone of a true die square, which can only be tried by the square and compasses. It represents the mind of man after a well spent life in acts of piety and devotion to God, and benevolence and good-will to man, which can only be tried by the square of God's Word, and the compass of his own conscience."

The Principal Point and the Original Signs are illustrative of Brotherly Love, Relief, and Truth; and of Temperance, Fortitude, Prudence, and Justice; all of which are moral duties emanating from that sacred Volume which is always spread open upon the Pedestal; and are copiously explained in the primitive lectures of masonry.

Another beautiful illustration of the poetry of the Order is found in its application of the virtues of silence or secresy, which is one of the distinguishing virtues of the masonic science, and is regularly enforced in the ordinary masonic lectures. Of all the arts which masons profess, the art of secresy particularly distinguishes them. Taciturnity is a proof of wisdom, and is allowed to be of the utmost importance in the different transactions of life. The best writers have declared it to be an art of inestimable value; and that it is agreeable to the Deity himself, may be easily conceived, from the glorious example which he gives, in concealing from mankind the secrets of his Providence. The wisest of men cannot penetrate into the arcana of heaven, nor can they divine to-day what to-morrow may bring forth.

A certain Key is also spoken of in the Prestonian Lectures, which ought always to hang in a brother's defence and never to lie to his prejudice; and the brethren are advised of the value of a tongue of good report, which ought always to treat a brother's character in his absence as tenderly as if he were present; and if unfortunately his irregularities should be such, that this cannot with propriety be done, to adopt the distinguishing virtues of the science.

This system was solemnly impressed upon the candidate in the mysteries of Egypt, whence originated the famous quinquennial silence of Pythagoras. The priests of Egypt were aware, if ever any set of men were acquainted with the maxim, that knowledge is power. The higher classes of the priesthood were extremely cautious how they communicated information to the younger and lower orders of the hierarchy; and these again were not less reserved in their intercourse with the rest of society. The numerous and dangerous ordeals through which the priests had to pass, and the long term of years allotted for their apprenticeship, sufficiently prove the truth of the statement which I have just been making. Every step by which the aspirant advanced, was preceded by a new trial of his patience, and a new proof of his fortitude. Before he passed into darkness, and when again he returned to the light, the object which still met his eyes, was the image of the god whose finger is on his lip. Silence and secresy were the first duties taught to the aspirant. He might listen, but he might not speak. If he heard a voice, it addressed him in the language of mystery. If he received information, it was conveyed to him through the medium of tropes and symbols.[3]

In Freemasonry this silence or secresy is urged on the brethren, that they may avoid speaking of a brother's faults; because human nature being imperfect, we are none of us free from errors of some kind; and therefore, as we are liable to censure ourselves, we should refrain from passing sentence upon others, that they may be actuated by a similar motive, and avoid all unfavourable reflections on our own conduct. It is an amiable principle, and highly beneficial to society; for what good can possibly arise from a public exposure of each other's foibles or miscarriages. If I err to-day, and my brother charitably passes it over, shall I expose the fault which he may commit to-morrow? Such a course would merit the severest reprobation. We have a rule, which, if universally observed, would produce more peace and happiness in the world, than, I am afraid, is to be found amongst mankind at present. It is a golden maxim, applicable to all times and occasions, and cannot possibly fail in its operation. It was delivered by the Divinity—taught in the gospel—recognized in Freemasonry—and is equally beneficial to all orders and descriptions of

[3] Drummond. Orig., vol. ii., p. 207.

men. These are the words. WHATSOEVER YOU WOULD THAT MEN SHOULD DO UNTO YOU, DO YE ALSO UNTO THEM.

This may be illustrated by a passage from the "Stray Leaves" of a Suffolk Rector. Speaking of an old soldier, whose latter years were spent in difficulties, he says: "Here was a man who unquestionably had spent the prime of his life in his country's service. He had carried her standard and had fought her battles. His blood had flowed freely in her cause. His adherence to her interests had cost him dear. Wounds, which neither skill nor time could heal, disabled him from exertion, and rendered life a burden. To acute bodily suffering positive privation was added. Who relieved him? His country? No. She left him to perish on a niggardly pension. Who succoured him? The great Duke, whose debt to the private soldier was so apparent and overwhelming? No. Who, then, aided the wounded and sinking soldier in his extremity? THE BROTHERHOOD—a secret band, if you will, but active—which requires no other recommendation, save desert, and no other stimulus than sorrow. And yet, *how little is it understood, and how strangely misrepresented.*"

If a brother, however, should grievously sin against the rules of the Institution or Society with which he is identified, we have another rule of conduct which is worthy of notice. "If thy brother shall trespass against thee, go and tell him his fault between thee and him alone. If he shall hear thee, thou hast gained thy brother. But if he will not hear thee, then take with thee one or two more, that in the mouth of two or three witnesses every word may be established. And if he shall neglect to hear them, tell it unto the assembly; but if he neglect to hear the assembly, let him be unto thee as an heathen man and a publican.[4] In the Book of Constitutions, (Private Lodges,) we find it provided, that if any brother behave in such a way as to disturb the harmony of the Lodge, he shall be thrice formally admonished by the Master; and if he persist in his irregular conduct, he shall be punished according to the Bye-Laws of that particular Lodge; or the case may be reported to higher

[4] Matt. xviii., 15–17.

masonic authority; but no Lodge shall exclude any member without giving him due notice of the charge preferred against him, and of the time appointed for its consideration. The proceedings against him are thus conducted with great caution and secresy. The erring brother must be privately admonished by the Master *thrice.* Some of these admonitions, it is hoped, may save him from exposure. If the two first should unhappily fail of their effect, the third is generally given in the presence of two or three confidential friends and brothers, as witnesses of the fact, and sometimes before the Lodge; and if this should also be disregarded, still mercy prevails—he is furnished with a further opportunity of repentance and amendment of life. The case may be referred to the P. G. Lodge, or the Board of General Purposes; and if he persist in his contumacy, he is punished by fine, suspension, or, in extreme cases, by expulsion. In the language above cited, he becomes, in our estimation, as an heathen man and a publican.

The Constitutions of the Grand Lodge of Massachusetts contains the following very judicious regulation on this subject: "The accusation shall be made in writing, under the signature of a Master Mason, and given in charge to the Secretary of the Lodge; who, under the direction of the Master, shall serve, or cause the accused to be served with, an attested copy of the charges, fourteen days at least previously to the time appointed for their examination, provided the residence of the accused shall be known, and shall be within the distance of fifty miles of the place where the Lodge having the matter in hand is located. If the residence of the accused be at a greater distance than fifty miles, then, and in that case, a summons to appear and show cause, forwarded to him by the mail or other conveyance, twenty days at least before the time of trial, shall be considered sufficient service. If his residence be out of the State, and unknown, the Lodge may proceed to examine the charges ex parte; but if known, a summons shall be sent to him by mail, or otherwise, sixty days at least before the time appointed for the examination; which shall be had in a Lodge specially notified and convened for the purpose, at which no visitors shall be admitted, except as counsel or witnesses. The accused may select any brother for his counsel, and witnesses shall testify, if masons, on their honour, as such. Hearsay evidence shall be excluded. The question—Is the accused guilty or not guilty? shall be put to each member of the Lodge, by name, commencing with the youngest. The answer shall be given standing, and in a distinct and audible manner, which shall be recorded by the Secretary. If the verdict be suspension or expulsion, an attested copy of the proceedings shall be sent up at the ensuing meeting of this Grand Lodge for examination and final action."

Again, the secresy of Freemasons is an effectual antidote to slander and defamation. These are vices of the most baleful kind, because they injure the credit of him who is the subject of false report, without benefiting the slanderer. Defamation is a crime of the blackest dye; it is founded in malice, propagated in hatred, and becomes the mischievous author of suspicion, envy, and all uncharitableness. Amongst numerous bodies of men, it must necessarily happen that characters will occasionally be found, how strictly soever the institutions of a society may guard against their introduction, who are base enough to pass unmerited censures on their brethren, even at the risk of sullying their own reputation; and the usual consequences will follow, if great care is not taken to crush this mischievous propensity in its bud, and check the rising evil before any fatal results are accomplished.

If not—if through favour, or fear, or timidity, or any other improper feeling in a Master of a Lodge, the necessary precautions are not adopted—if the enjoined admonitions be deferred from time to time, or postponed ad infinitum, he will soon find his Lodge in a state of insubordination and misrule, which will be highly discreditable to himself, and inflict a great portion of evil on the community which he governs.

And more than this; society will suffer from the bad example thus exhibited; for the disorders of a Lodge, like those of a city built upon a hill, cannot be concealed; and the most disastrous effects may possibly ensue from the misconduct of a member, augmented and strengthened by the discreditable connivance of the Master, whose duty it was to discountenance every attempt to violate the institutes of Masonry, amongst which the recommendations to avoid slander occupy a prominent situation. For what good can be expected to arise out of whisperings, backbitings, debates, strife, variance, emulations, anger, and evil speaking? A high authority proclaims that if any man seem to be religious and bridleth not his tongue, but deceiveth his own heart, this man's religion is vain.

It was an excellent regulation of our own Grand Lodge in the last century, that when any brother was proposed to join a Lodge, or any candidate to be initiated, and it should appear upon casting up the ballot, that he was rejected; it was absolutely forbidden that any member or visiting brother should discover, by any means whatsoever, who those members were that opposed his election, under the penalty of such member being forever expelled from the Lodge, and if a visiting brother, of his being never more admitted as a visitor, or allowed to become a member; and immediately after a negative passes on any persons being proposed, the Master shall cause this law to be read, that no brother present may plead ignorance.

This law ought to be revived, for a talebearer is unworthy of a place amongst honest men. He is a despicable character, and ought to be avoided. He enters a Lodge—listens to everything that is said, and reports it abroad with numerous exaggerations, and generally under a pretended seal of secresy—as if those on whom he obtrudes his information care anything about his injunctions not to repeat the calumny. The slander spreads far and wide, and, like a secret poison, becomes incurable before the injured person knows anything about it. This, therefore, may justly be accounted one of the most cruel wounds inflicted by a tongue of evil report; for it undermines society, and frequently robs families of their peace, and innocent persons of their good name. It separateth chief friends; and, therefore, a tongue that is given to this wicked practice, may be properly said to be set on fire of hell.

For instance, a neighbour has acted indiscreetly. The story is conveyed from ear to ear. It is carried from house to house. It is the topic of every circle. The evil-speaker hears the tale with rapture, and with rapture relates it. He enlarges upon the enormity of the crime; he lashes it with severity; he loads the actor of it with the harshest epithets with which the language is able to supply him. Is he ashamed of his want of lenity and mercy? Does he blush before his conscience when he retires into himself, and looks to the heap of stones, and hard ones, too, which he has thrown? When he sinks upon his pillow, will the recollection of the words that have gone from him allow him to sleep? His sleep is as sound as yours. He flatters himself that he is actuated solely by a virtuous abhorrence of iniquity.[5]

Let every Master of a Lodge, therefore, when he hears an unfavourable report of any individual brother, which he has reason to think false or exaggerated, consider *himself* as an injured party, and bound by the duties of his office to do justice to a calumniated friend, who may, perhaps, be unconscious that his reputation has been assailed.

It would be easy to proceed much further in illustration of the poetry and philosophy of Freemasonry, but it will be unnecessary, as enough has been already said to show the nature of its working, and the effect which such a system is sure to produce upon a great majority of the members. And if carried into general practice, cannot fail to insure the most beneficial results to society, by the admixture of even that small portion who have received the benefit of masonic instruction; because it is the sentence of one wiser than man, that "a little leaven leaveneth the whole lump."

The superficial mode which is at present used by many of our country brethren of conducting a lodge, is totally inefficient. And it can scarcely be otherwise, when only two or three hours in every month are devoted to the purposes of Masonry; and out of which, the routine business of management—the propositions, ballotting, initiations, passings, raisings, and desultory motions, occupy so much time, that little remains for the purpose of pursuing the studies necessary to a complete knowledge of the science. At best, the Lodge Lectures are too circumscribed for a course of general instruction; and yet they are quite as explanatory as

[5] Fawcett's Sermons at the Old Jewry, vol. 1., ser. 9.

the nature of the circumstances will admit; for in the limited portion of time which can be assigned to their delivery, it would require almost the whole twelve months from festival to festival to go deliberately through the entire lectures of the three degrees. For this reason many Lodges confine themselves to the first three or four sections of the E. A.˙. P. Lecture, and seldom touch on the other two, except at passings and raisings; some are content with a simple explanation of the Floor Cloth or Tracing Board; while others seldom venture beyond the Qualification Questions!

Now it will be readily admitted that Freemasonry, as it ought to be, is invested with higher views and more interesting and useful objects of contemplation. By the principle of association, and a mutual interchange of sentiments, it inculcates brotherly love among all mankind; it tends to soften the harshness of an exclusive or sectarian feeling towards those who differ from us in our views of religion and politics, although it allows of no discussions in either the one or the other; it suppresses the attachment to class, which is the bane of all other institutions; and by the purity of its sentiments, it harmonizes the mind, ameliorates the disposition, and produces that genuine feeling of benevolence and Christian charity which "suffereth long and is kind; which envieth not, vaunteth not itself, is not puffed up, doth not behave itself unseemly, seeketh not her own, is not easily provoked, thinketh no evil, rejoiceth not in iniquity, but rejoiceth in the truth; beareth all things, believeth all things, endureth all things."[6]

The above principles are almost exclusively Christian, and afford ample evidence that a corroboration of the moral precepts of Freemasonry will be found in the Gospel of Christ. A talented Brother, with whom I have had an extensive correspondence on the subject of masonry, writes thus:—"Your hypothesis that the Lectures of Masonry, as now authorized by the Grand Lodge, are intended to enforce the great truth of Christianity, is undoubtedly correct. And as they were framed by a clergyman of the Church of England, less was scarcely to be expected. But I contend that all

[6] Cor. xiii. 4—7.

allusions to Christianity are interpolations in the system. In a mere Blue Lodge, which I maintain to have been originally restricted to working masons, with very few exceptions, nothing more was required than a moral explanation of the Bible, Square, Compasses, Level, and Plumb. In Scotland the three first degrees were considered to be confined almost entirely to science, and the correct definition of masonry is—A science founded on Geometry, Mathematics, and Astronomy. And accordingly the top of the Master's Rod of Office is surmounted by a triangular spear head, on which are the letters G. M. A. The Scotch masons consider the moral explanation, if obvious and simple, to be proper, but refer all deep and mystical topics to a superior degree. In short they allow of no allusion to the New Testament, nor to anything in the Old Testament after the book of Kings and Chronicles, referring to the Temple of Solomon; and there must be no anachronism. All after the building of the Temple, are topics that cannot be touched on until we arrive at the Royal Order of H. R. D. M.; and therefore it is not en règle to refer to the chief corner stone till the appearance of a Christian degree. Faith, Hope, and Charity have no business in the lectures of the Blue degrees; unless, indeed, we are to abandon our claims to antiquity, and admit that Freemasonry is a fabrication, invented at some recent period subsequently to the crucifixion of Christ."

It will be observed, however, that Christian allusions abound in the lectures of masonry long before Dr. Hemming remodelled them in 1814. They exist copiously in the very earliest masonic manuscripts known; which Mr. Halliwell pronounces to be a production of the 14th century; while others consider them to be coeval with the time of Athelstone. Christian references are also found in the first lectures authorized by our own Grand Lodge in 1720. In fact Dr. Hemming, so far from introducing into his formula any new allusions to our most holy faith, actually expunged some of those which were in use before his time.

The first lectures after the revival, when it was arranged that "the privileges of masonry should no longer be restricted to operative masons, but extend to men of various professions, provided they were regularly approved and initiated into the Order," contained many Christian references, which were gradually increased in every successive arrangement, until Hutchinson, about the year 1784, interpreted the third degree as being exclusively Christian. Now although I cannot subscribe to this view of the case, it shows at least the feelings of our brethren of the last century on this particular subject; and it is my deliberate opinion, that it even the group of symbols which form the subject of this volume, was struck out of Freemasonry, and it forms chiefly an illustration of the first degree, the system would be so thoroughly impoverished that it would fail to interest the mind even of an indifferent enquirer; while the more talented candidate would take leave of us on the threshold, and consider the charges of frivolity and uselessness, which have be enpreferred by our enemies, to be amply confirmed.

That this can never happen in masonry as it is at present constructed, will be shown by the evidence of my friend Bro. Tucker, P. G. M. for Dorset; who, said, in his speech at Weymouth, 1846:—"The whole of our proceedings stamp the institution of Freemasonry with a character, divine in its origin, holy in its purposes, and conducive to the best interests of man. We will not enquire how far it may be supposed to be allied in form to the ancient Druid in his rites and mysteries, or in the erection of his temple, nor to the refined philosophy of the early Greek, or the dark and mysterious knowledge of the Egyptian hieroglyphic; neither will we consider how far we are warranted in applying the use of familiar masonic terms to the ancient patriarchs to whom came the divine message to man in all the power and terrific grandeur of heavenly majesty, as well as in the sweetness of divine love, in the still small voice of mercy; but we will take it on its own merits, as founded on the Word of God, as the guide of our days, and setting before us the hope of eternal life;—an institution equally apart from bigotry and fanaticism, teaching us to walk in the good old paths of our forefathers; to do justly, to love mercy, and to walk humbly with God, being also heir with them of the same promises, and endeavouring to draw all mankind of every clime, colour, and religion, within the circle, to that point from which a master mason cannot materially err."

LECTURE III.

Epistle Dedicatory

TO

BRO. R. GRAVES,	W. M.
— E. D. SMITH,	S. W.
— R. COSTA,	J. W.
— W. L. WRIGHT,	P. M. & TREA.
— G. CHANCE,	SEC.
— M. COSTA,	S. D.
— W. L. W. APLIN,	J. D.
— L. CAMPANILE,	STEWARD.
— R. SPENCER,	P. M. & M. C.
— Z. WATKINS,	PAST MASTERS,
— J. N. BAINBRIDGE,	PAST MASTERS,
— S. BRIZZI,	PAST MASTERS,
— E. MULLINS,	PAST MASTERS,
— J. WHITMORE,	PAST MASTERS,

Of the Bank of England Lodge, No. 329, *London.*

MY DEAR BRETHREN,

I have much pleasure in dedicating to you the following observations on the Lectures on Masonry, as they were arranged at the Union in 1813, and directed to be used in all the private lodges under the Grand Lodge of England; and am right glad that a public opportunity has occurred of acknowledging the kindness which you have extended to me on several occasions, and of expressing the gratification I have ever felt in being associated, as an honorary member of the lodge, with sc many eminent men, whose zeal and services in the cause of masonry have justly excited the approbation of the fraternity, and placed them high in the estimation of the wise and good.

It will be needless to repeat my opinion of the Order which we venerate and profess. It is well known that I have bestowed much attention on the subject both as a theoretical and a practical science, and the results of my enquiries are before you.

The benefits arising from a competent knowledge of the poetry and philosophy of Freemasonry are open to every studious person, and may be easily attained by a proper exercise of the mental faculties. It is by care and industry that every earthly good is secured. The Freemason, therefore, who expects to reap any intellectual advantages from the Order, must study its principles with diligence and assiduity, as you have done, else he will fail in the attempt.

A true knowledge of the science will not be acquired by indolence and apathy, nor by a mere acquisition of its signs, and tokens, and technicalities. These are but the keys to our treasure. The cabinet must be opened, and its contents examined carefully, and with an ardent desire to profit by the materials which are deposited there.

If a brother be desirous of becoming useful to the science of Freemasonry, he will not be content with a mere superficial knowledge of the externals, but will examine its esoteric secrets with the feelings of an enthusiast; and by bringing forth its latent virtues into view, will himself reap a full share of the blessings which it is so well calculated to confer on society at large.

It is by the practice of such a judicious course of study that the brethren of 329 have distinguished themselves; and the acknowledgment of such a belief will not be thought presumptuous or inappropriate, when avowed by one who has the greatest pleasure in thus subscribing himself,

My dear Brethren,

Your obliged and faithful

Servant and Brother,

GEO. OLIVER, D. D.,

Hon. Member of the Lodge.

SCOPWICK VICARAGE,

August 1, 1849.

Lecture the Third.

A few observations on the Lodge Lectures, with the means of acquiring a knowledge of them.

"Bro. Lane said he had derived much pleasure and instruction from that source of knowledge which is contained in published works on Masonry. Those who know anything of the Continent, know that large collections of books exist in masonic societies there, and, that many valuable works were in this country, which the library, if established, might some day hope to possess. He had collected several rare and costly works on Masonry, valuable, even in the places where they were published and best known, for their scarcity; these he intended to present if the library were established, and carried on under regulations that were satisfactory to him."—*Debate in Grand Lodge on the formation of a Library and Museum.*

The Lectures of Freemasonry teach—and if they taught nothing else, their value would be incalculable—that it is only by the practice of the relative and social duties of life that our present condition can be benefitted, or even maintained. The discharge of these permanent obligations, will make good masters, as well as good servants; good magistrates, as well as good subjects; kind husbands, and faithful wives; for all have duties to perform, the absence of any one of which would break the chain of social relations, and destroy the peace and happiness of those who are unfortunately placed under its influence. A vicious parent, by evil example, will demoralize the principles of his offspring; and the consequences may be transmitted for years to come; as is the case with some physical peculiarities and blemishes; whence arises the bad character which we frequently find attached to particular families; and adheres to them and their descendants, who inherit their mischievous propensities, sometimes through many generations.

The proposition will hold good when applied to a masonic lodge. If the Master be addicted to intemperance, the brethren will eagerly imitate the example, and plead it as an excuse for their own irregularities. But such a plea, though it may satisfy the conscience of an offending person, will avail him nothing in mitigation of the punishment which is due to his crime, whatever it may be, either in this world or in that which is to come. Would it be accounted a valid excuse in a court of justice, for a prisoner to urge the legality of his having committed a murder or a robbery, because others had done the same, or because they persuaded him to do it? Or will the laws of Masonry be invalidated, if an erring brother should plead—"I only imitated the example which had been set by the W. M. when I got intoxicated, or slandered a fellow creature; and therefore, he is the transgressor and not I." He might with equal justice blame the genial influence of the sun because it brings poisonous, as well as salutiferous, herbs to maturity.

In the Book of Constitutions this is guarded against by a series of judicious regulations which can neither be evaded nor misunderstood. Indeed, the first lesson which is taught to a candidate is, the necessity of a strict adherence to his relative and social duties. And to give this the greater effect, it is directed to be done by the Master, in a Charge which he is enjoined to deliver at every initiation. In this Charge the following beautiful passage occurs. "As a citizen of the world, I am next to enjoin you to be exemplary in the discharge of your civil duties, by never proposing, or at all countenancing, any act that may have a tendency to subvert the peace and good order of society; by paying due obedience to the laws of any state which may for a time become the place of your residence, or afford you its protection; and, above all, by never losing sight of the allegiance due to the sovereign of your native land; ever remembering that Nature has implanted in your breast a sacred indissoluble attachment to that country from which you derived your birth and infant nurture." Indeed, the same Charge declares that, the practice of social and moral virtue constitutes the solid foundation on which Freemasonry rests. And this view is borne out by the general teaching of the Lodge.

A knowledge of the Lectures of Masonry is accomplished by a system of mutual instruction which encourages and rewards industry. Indolence is, indeed, the parent of every vice. "If you ask me," says Lavater, —"if you ask me which is the real hereditary sin of human nature, do you imagine I shall answer pride, or luxury, or ambition, or egotism? No; I shall say Indolence; who conquers indolence will conquer all the rest." It has been justly remarked that if the mind of man be not employed in good, it will be employed in evil. And hence spring the numerous crimes which deform society, and lead to a painful and ignominious death.

The sagacious Greeks saw this in its true light, and their legislators provided against it by the introduction of judicious laws. Solon, as well as Draco, began with childhood, and provided for the good conduct of the future citizen by assigning masters adapted to the character and talents of the children; and especial care was taken that no evil communications should contaminate their minds. A court of justice was appointed to superintend the process of education; and if any improper person obtruded himself unnecessarily into the presence of the children, he was punished with death. When arrived at maturity, the school was changed for the gymnasium; and they were still under the superintendence of the law, that the dangers of evil example might be avoided, and purity of manners secured. After this, rewards were assigned to virtue, and punishment to vice.

A similar plan is pursued in a mason's lodge. The system of lecturing which is there used, if industriously and faithfully pursued, will produce the same effect, by extinguishing idleness, and promoting a spirit of enquiry and thought. Every person becomes desirous of excelling; and this induces an earnest attention and application to the business in hand. The offices of the lodge are open to none but such as, by diligent reflection, have formed their minds to a habit of reasoning, which is the forerunner of knowledge, and enables them to exchange the character of pupils for that of teachers. The judicious division of the Lectures into sections and clauses, affords ample facilities for improvement; and by acquiring a competent knowledge of the parts; by conquering the graduated steps in detail; the tyro soon becomes master of the whole; and the excellency to which he thus visibly approaches, recommends him to the notice and applause of the brethren.

The knowledge thus acquired is a species of wealth which is endurable, and cannot be taken away. When the city of Megara was captured by Demetrius, and the soldiers were about to plunder it, the Athenians, by a strong intercession, prevailed on the general to be satisfied with the expulsion of the garrison. There was residing in the city at that time a celebrated philosopher whose name was Stilpo. Demetrius sought him out, and asked him if the soldiers had taken anything from him. He answered, "no, none of them wanted to steal my knowledge."

A habit of systematic regularity being once attained by the practice of the lodge, it soon becomes characteristic of the man; and this principle, judiciously exercised, will lead him to eminence, whatever may be the station which he occupies in the world. A heathen poet could tell us that idleness is the prolific parent of all vice.

> Quæritur Ægystus quare sit factus adulter;
> In promptu causu est; desidiosus erat.

On the other hand, perseverance is always successful; for that which is attributed to misfortune, may often be the effect of imprudence or inattention. How frequently do we hear complaints from indolent men, that their time is so fully occupied in providing for the necessities of their families, that they have no leisure for speculative pursuits, when in fact there are more hours wasted in frivolities by such men than would serve to make them masters of all the arts and sciences, if they were properly applied. When Philip, King of Macedon, invited Dionysius the younger to dine with him at Corinth, he felt an inclination to deride the father of his royal guest, because he had blended the characters of prince and poet, and had employed his leisure in writing odes and tragedies. "How could the King find leisure," said Philip, "to write these trifles?" Dionysius answered, "in those hours which you and I spend in drunkenness and debauchery."

By the practice of industry, even during the short period employed by the master in delivering his periodical instructions, any Brother may improve his mind by acquiring a competent knowledge of the Lodge Lectures; and they will abundantly reward his labours, by leading him to regard the works of creation not merely with the eye of a philosopher, but with that of a Christian. They will teach him to look from Nature up to Nature's God, as displayed in his glorious works in the starry firmament, which every mason who is desirous of becoming perfect in the art should study with attention, as they display the wonders of his handy work. The canopy of the Lodge is an open book where he may read the tokens of power and magnificence which display the absolute perfection of T G A O T U. The annual recurrence of vegetation and decay affords striking indications of his powerful Hand, but the beauties with which he hath decorated the heavens, are evident manifestations of his supremacy, still more sublimely apparent. They harmonize with his Holy Word, and contain incontrovertible proofs of its truth; and the Master of a Lodge who omits to draw the attention of the brethren to these august phenomena, is deficient in his duties, and fails to make the science of Freemasonry subservient to the great end for which it is principally designed—the glory of God and the welfare of man.

The true mason will look with sentiments of awe and veneration on these and other great works which are open to his observation, although not, perhaps, specifically mentioned in the lectures. I refer to the treasures as well as the terrors which the earth contains within its bowels; minerals and metals; boiling springs and burning mountains; earthquakes and simoons, pestilence and famine. All these, if judiciously introduced as an illustration of certain portions of the lectures, will prove highly interesting to the brethren, and enable the intelligent Master to refer them severally to the power of the Most High. "For it is the Lord that commandeth the waters; it is the glorious God that maketh the thunder; it is the Lord that ruleth the sea; the voice of Jehovah is mighty in operation; the voice of Jehovah is a glorious voice. It breaketh the cedar trees; yea, it breaketh the cedars of Libanus. He maketh them also to skip like a calf, Libanus also, and Sirion, like a young unicorn. The voice of Jehovah divideth the

the flames of fire; the voice of Jehovah shaketh the wilderness; yea, the Lord shaketh the wilderness of Cades."[1]

All our scientific disquisitions are invested with the same tendency. They serve to make us wiser and better men; and if they fail to do so, the fault is not in the institution, but in the indifference of the recipient to the real object and design of masonic teaching.

It will be readily admitted that the details of Freemasonry are somewhat faulty, and their amendment would tend to increase not only the benefits but the popularity of the Order. In the United States these details are carried out with much better effect than in some of our country Lodges; the consequence of which is, that the Order is so universally and widely diffused throughout every class of the community in that Republic, as to constitute almost a national establishment. This is owing, however, in some measure to the amicable antagonism and social intercourse of its numerous independent Grand Lodges, which being placed amidst the private Lodges under their individual jurisdiction, they are enabled to superintend the working, to restrain disorders, and to apply an instant and effectual remedy for any irregularity which may spring up amongst them.

It is a question, which the fraternity may answer, whether, if every province in England had its independent Grand Lodge, masonry would not be more prosperous, more regarded, and more abundantly useful. As this, however, is a consummation which can scarcely be reduced to practice, we must consider whether certain improvements may not lawfully be accomplished without it. And for this purpose we will again refer to the usage of the United States of America. One great cause of the overwhelming influence of transatlantic masonry, is the extensive encouragement given by its Grand Lodges to publications on masonic subjects. Their language is unanimous on this point. The Grand Lodge of New York, in its printed transactions, thus expresses its opinion. "In reference to the several masonic periodicals named by our D. G. Master, if judiciously conducted, as your committee entertain no

[1] Psalm xxix., 2—7.

doubt they will be, they are calculated to accomplish a vast amount of good, by diffusing more extensively those sound, moral, and benevolent principles, which so eminently characterize this venerable institution; we therefore recommend those publications to the liberal patronage of the fraternity."

And again in the same document, we find the following clause, "In relation to the recommendation of the D. G. Master, desiring some action of the Grand Lodge with reference to Lectures to be delivered in the several Lodge-rooms, not only on the principles of masonry, but on the Arts and Sciences, embracing any or all such subjects as shall improve the moral and intellectual powers, and qualify the brethren for greater usefulness in the several spheres in which they move, rendering them, in an eminent sense, the *lights* of masonry; we are of opinion that the adoption of such a practice would be attended with the happiest results, and add much to the interest of fraternal communications. The masonic institution is appropriately a school of the Arts and Sciences, as well as the moral virtues; and therefore the Grand Lodge recommend in the strongest language, the adoption of the above specified course of instruction by Lectures on the practical, as well as the moral and mental sciences, in each of the subordinate Lodges. This whole matter appears to be one of deep interest, and if judiciously conducted by brethren competent to the undertaking, will not only be productive of great good to individual members, but to those communities where such lodges are established."

The Grand Lodge of New Hampshire is equally explicit. Its Grand Master in 1843 thus promulgated his sentiments from the throne: "You will permit me, brethren, to bring before you, for your countenance and support, the Freemason's Monthly Magazine, published in Boston, Massachusetts, under the editorial charge of our valued brother, R. W. Charles W. Moore, Secretary of the G. L. of Massachusetts, and former publisher and editor of the Masonic Mirror, which was suspended during the dark years when the anti-masonic party of that State followed up persecution on the rights of our ancient and honourable institution, with a zeal that could hardly be commendable, if used for a better purpose. This work is conducted with ability and interest. It is the fruit of reflection and study; animated by a spirit that breathes love to man, and expresses in clear tones the faith of an institution that will outride all persecution; implanting in the heart of the initiated that charity which suffereth long and is kind. To such a work, conducted on the true principles of Freemasonry, which cannot fail to further the progress of the Craft,—I trust, and doubt not, you will give your support." This avowal and recommendation was warmly and eloquently advocated by several members of the Grand Lodge, who bore ample testimony to the high and exalted character of that publication; and it was unanimously recommended to "the Members of the Fraternity every where, as worthy and highly deserving their patronage and support."

Nor do we find throughout the United States an adverse opinion on this subject. Indeed, the several Grand Lodges attach so much importance to printed disquisitions which may be accessible to the brethren, and conduce to promote that degree of uniformity which is so desirable amongst the Craft, that at a general Convention of all the Grand Lodges in the States, holden at Baltimore in Maryland, May, 1843, a Committee was appointed to prepare and publish at an early day a text book, to be called the Masonic Trestle Board, embracing three distinct, full, and complete Masonic Carpets, illustrative of the three degrees of ancient craft masonry; together with the ceremonies of consecrations, dedications, and installations; the laying of corner stones of public edifices; the funeral service, and order of processions; to which shall be added the charges, prayers, and exhortations, and the selections from Scripture, appropriate and proper for Lodge service. The Committee further reported, that they deem it expedient that a work be published to contain archeological researches into the history of the Fraternity in the various nations of the world. In compliance with a formal order to that effect, the Lecture on the Trestle Board has been printed, and is now before the public.

In England there is an influential party whose study it is to discourage all scientific disquisitions connected with Freemasonry. Such a course, if persisted in, would throw us back upon the dark ages, and furnish our adversaries with a weapon which they would not fail to wield with fearful effect. But happily we have a sufficiency of liberal minded brethren who are anxious to place the beauties of masonic benevolence, and the excellencies of masonic science fairly before the public, that its light may shine before men, to the glory of T G A O T U who is in heaven.

These two sections entertain very different opinions on the best means of promoting and cementing the general interests of the Craft. The former think it inexpedient to allow any alterations to be made in the system; lest, under the idea of improvement, innovations should creep in, which might, in process of time, change the very essence of the institution, and gradually deprive it of those characteristics which are considered to be its greatest ornaments. The other party, who are also numerous and influential, are of opinion that masonry ought to keep pace with all other scientific improvements; and that in the rapid progress of mental enlightenment, which distinguishes the present era, if this Order alone should remain stationary, and take no part in forwarding the march of intellect in its own peculiar sphere, it will forfeit its claim to public notice or approbation.

In the present state of intellectual improvement, men do not meet together for the insane purpose of hearing repetitions of truisms with which they are already acquainted. Their minds reach forward to something new. They will not consent to retrograde, nor are they satisfied with remaining stationary. Time is considered too valuable to be wasted without actual improvement; and it is by the exercise of the intellect that it is strengthened and rendered capable of renewed exertion. To Freemasonry, as in all other human pursuits, the onward principle must be applied, if we would make it applicable to the poetry and philosophy of life; or the paralyzing question, *cui bono?* will be surrounded with difficulties that, in the end, will be found inextricable. The time is drawing near when the investigations of masonry must be carried on in common. Every member will demand his share of the work. The W. M. will be the Moderator to preserve the unanimity of the proceedings; and his task of instructing and improving the brethren in masonry, will admit of a participation with other talented brethren, who have had as much experience as himself. By such a course the Order will be ennobled, and will claim and receive the respect to which it is fairly entitled.

The former section of the Craft entertain a great aversion to publications on the subject of masonry, and discountenance them by every means in their power, under an impression that such writings, although exclusively confined to the philosophy and poetry of the Order, are calculated to do a great deal of mischief by enlightening the profane on subjects of which they ought to remain for ever in ignorance, except their knowledge be acquired through the legitimate medium of initiation.

But if nothing were lawful but what is absolutely necessary, ours would be but a miserable world to live in. Literary talent would be circumscribed within a very narrow compass; science would be consigned to oblivion; the fine arts be suffered to decay; and we should return to the state, almost savage, of the primitive inhabitants of this island, who dwelt in dens, and caves, and wretched hovels, and according to Dio Nicæus, would bury themselves in bogs up to their neck, and continue in that state for many days together without sustenance; and retiring from thence they would take shelter in the woods, and feed upon the bark and roots of trees. Instead of clothing, their bodies were tattooed with divers figures of animals and plants; living continually at war with their neighbours, and offering their prisoners in sacrifice to the gods. The above objection, therefore, is not of sufficient weight to counterbalance the benefits which are derivable from masonic investigations.

For these reasons, the latter class of our brethren entertain a reasonable opinion that Freemasonry ought not to linger behind any other scientific institution; but in its onward progress ought to run parallel, at the least, with the very foremost, towards the goal of perfection. For this purpose they give their full sanction and encouragement to printed disquisitions on the general principles of the Order, because they consider such productions to be eminently calculated to carry out the scientific and benevolent designs of masonry, and to cause those who have been most bitter and active in their hostility, to entertain more modified views of the institution, and even to solicit the honour of initiation, that they may become acquainted with its real tendency and design.

The former would prefer the alternative, that errors and imperfections should eternally deform the institution, rather than see them dragged forth and exterminated by legislative enactment; and that silence on the subject will make the world believe Freemasonry to be perfect, and unimprovable even in the minutest particular. But mankind are not so easily deceived. They know very well that no human institution is perfect; and their lynx eye is too penetrating, notwithstanding all our secresy and all our care, to allow our imperfections to escape their notice. If, therefore, we wish Freemasonry to be publicly esteemed as a popular establishment, let us boldly apply the actual cautery, and expunge every questionable doctrine and practice from the system; for a cure cannot reasonably be expected, unless we discover the cause of the disease.

Amongst the latter class we find many successive Grand Lodges from the time of the great revival in 1717, as is evidenced by the organic changes to which they have given a decided and unequivocal sanction. As for instance, the gradual increase in the names and number of the officers of Lodges. Originally these consisted of three only. In 1721, a Deputy Grand Master was first appointed. In 1722, the office of Secretary was instituted; and this was succeeded in the following year by the nomination of Stewards; but it was not till 1730 that the office of a Treasurer was added to the list. In 1732 a Sword Bearer; but the office of a Deacon was unknown till the very latter end of the century. These were succeeded by Grand Chaplains, Architects, Portrait Painters, and, after the reunion in 1813, by an officer called Pro Grand Master which, however, appears to be considered necessary only when a Prince of the blood royal is on the throne.

These are all changes in the original system, and were introduced for the improvement of the Order, under the following law of the first Grand Lodge; "that any Grand Lodge duly met, has a power to amend or explain any of the printed regulations in the Book of Constitutions, while they break not in upon the ancient rules of the fraternity."

It appears, then, from the above authority, that alterations, not affecting the Landmarks or fundamental principles of masonry, may be lawfully made under the sanction of the Grand Lodge; and also that such alterations have, from time to time, been considered necessary by that body, to meet the requirements of an improved state of society.

The question then arises, what are the Landmarks of Masonry, and to what do they refer? This has never been clearly defined. I have already recorded my opinion on the Historical Landmarks, in a voluminous work, expressly written for their elucidation; but it will be remarked that these are only *the Landmarks of the Lectures*, which, though practised by the fraternity under the above high sanction, have been almost entirely introduced into the system since the period of revival in 1717. There are other Landmarks in the ancient institution of Freemasonry, which have remained untouched in that publication; and it is not unanimously agreed to what they may be confined.

Some restrict them to the O B, Signs, Tokens, and Words. Others include the ceremonies of initiation, passing, and raising; and the form, dimensions, and support; the ground, situation, and covering; the ornaments, furniture, and jewels of a Lodge, or their characteristic symbols. Some think that the Order has no Landmarks beyond its peculiar secrets. And the Rev. Salem Town, long the Grand Chaplain of the State of New York, whose book on the Speculative Masonry was published under the sanction of the highest masonic authorities in the country, expressly declares that *our leading tenets are no secrets.* And again, "by a full and fair exposition of our great leading principles, we betray no secrets." Colonel Stone, in his Letters on Masonry and Antimasonry, says, "from the period at which I reached the summit of what is called ancient masonry, I have held but one opinion in relation to masonic secrets; and in that opinion I have always found my intelligent brethren ready to concur. It was this;—that the essential secrets of masonry, consisted in nothing more than the signs, grips, pass-words and tokens, to preserve the society from the inroads of impostors; together with certain symbolical emblems, the technical terms appertaining to which served as a sort of universal language, by which the members of the fraternity could distinguish each other in all places and countries where lodges were instituted, and conducted like those of the United States."[2] Another American writer affirms that "the secrets of masonry are her signs, words, and tokens; these the oath regards and no more. The common language of masons, in conversation on the subject of masonry, is a proof that this is the opinion of the fraternity in respect to the application of the oaths."[3]

If we adopt any of the above views of the subject, it will lead to a full conviction that some of the Landmarks have sustained considerable modifications, in order to adapt them to the improvements in science and morals which have distinguished the period when they were introduced. For instance, it is generally supposed that the O B is a Landmark. The Ex-President Adams, in leading the crusade against Freemasonry in the United States, A. D. 1834, which he hoped would elevate him to the presidency, calls it the chief Landmark of masonry, and that on which the very existence of the Order depends. And he adds dictatorially; "the whole cause between Masonry and Antimasonry, now upon trial before the tribunal of public opinion, is concentrated in one single act. Let a single Lodge resolve that they will cease to administer the O B, and that Lodge is dissolved Let the whole Order resolve that it shall no longer be administered, and the Order is dissolved; for the abolition of the O B necessarily imports the extinction of all the others."

This is an extreme opinion; but there are many amongst ourselves who entertain a similar belief. Let us, then, enquire whether any alterations have been permitted on this vital point. There are very cogent reasons for believing that primitive Freemasonry had but one O B for all the three degrees, which was short, expressive, and compact; and the penalty has been handed to our own times as an unalterable landmark. It was in this form before the year 1500, as appears from the old masonic manuscript which has been published by Haliwell, "A good trwe othe ehe ther swere to hys mayster

[2] Letter vii. [3] Ward's Freemasonry, p. 144.

and hys felows that ben there; that he will be stedefast and trwe also, to all thys ordynance, whersever he go, and to hys lyge lord the kinge, to be trwe to hym, over alle thynge. And alle these poyntes hyr before to hem thou most nede be y swore." The points here referred to were condensed by Desaguliers and his colleagues Payne, Anderson, Sayer, Morrice and others in the O B of 1720.

At present every degree has its separate O B, with penalties modelled on the original specimen. But even the first O B has sustained several alterations under the sanction of different Grand Lodges; and at the reunion under the two Grand Masters, the Dukes of Kent and Sussex, when a new arrangement of the Lectures was entrusted to the Lodge of Reconciliation by the United Grand Lodge, the ancient penalty was modified, and its construction changed from a physical to a moral punishment.

I would not have it understood that I disapprove of the alteration; although there are masons who consider it as the removal of an ancient Landmark, because I belong to that class who think that masonry, being a progressive science, is susceptible of improvement in accordance with the temper and intelligence of the age, without trenching on established Landmarks. I agree with Grand Master Tannehill when he says, "the Landmarks of the Order have existed through unnumbered ages, if not precisely in their present form, at least without any essential variation, although they have been handed down from age to age by oral tradition. The progress of society, the various changes that have taken place in the political, religious, and moral condition of mankind, have probably introduced various modifications in the forms and ceremonies of the Order; still its fundamental principles, and those characteristics which distinguish it from other human institutions, remain the same; so that by its symbolic language, a mason of one country is readily recognized and acknowledged in another. To preserve these Landmarks, and transmit them to our successors, is a duty we owe to posterity, and of which we cannot be acquitted so long as moral obligation has any force."[4]

[4] American Masonic Register, vol. iv. p. 1.

The alteration of the Master's word is another instance of the discretionary power which is vested in the Grand Lodge, of authorizing organic changes; for although not expunged, it was translated from the third degree to the Royal Arch by the Grand Lodge, of England *after the middle of the last century*, and a new word substituted in its place. Before that period its masonic meaning was explained by the words, "the Grand Architect and Contriver of the Universe, or he that was taken up to the top of the pinnacle of the holy Temple." Now, as no one was ever taken to the top of the pinnacle of the holy Temple but Christ, or the second person in the Trinity, there can be no mistake as to the meaning which our ancient brethren assigned to that sacred and sublime WORD.

This construction was kept pre-eminently before the fraternity in every code of lectures which the Grand Lodge thought it expedient, as society advanced in intelligence, to recommend to the practice of the subordinate lodges. A series of types were first introduced; then they were explained as being applicable to the Messiah; and an illustration was appended explanatory of the five great points of his birth, life, death, resurrection, and ascension. The herald and the beloved disciple were constituted the two great parallels of the Order, and symbolized by the figure of a circle, point, and parallel lines, which I have already, in a little work, devoted expressly to the subject, examined in detail; and to which I would refer the curious reader for further information, respecting these two presumed patrons of masonry. The three great virtues of Christianity were embodied in another emblem on the same road to heaven; and which, as the authorized lectures expressed it, "by walking according to our masonic profession, will bring us to that blessed mansion above where the just exist in perfect bliss to all eternity; where we shall be eternally happy with God, the Grand Geometrician of the Universe, whose only Son died for us, and rose again that we might be justified through faith in his most precious blood."

Many of the above illustrations were expunged by Dr. Hemming and his associates in the Lodge of Reconciliation, from the revised lectures; Moses and Solomon were substituted as the two masonic parallels, and T G A O T U was referred to God the Father instead of God the Son; forgetting, as Bishop Horsley observes, that "Christ, the Deliverer, whose coming was announced by the prophet Malachi, was no other than the JEHOVAH of the Old Testament. Jehovah by his angels delivered the Israelites from their Egyptian bondage; and the same Jehovah came in person to his Temple, to effect the greater and more general deliverance, of which the former was an imperfect type."

The above changes were made under the idea that masonry, being cosmopolite, ought not to entertain any peculiar religious tenets, lest, instead of being based on the broad foundation of universality, it should dwindle into sectarianism. But, without reminding you, that so far from being a religious sect, Christianity, if we are to believe the Jewish or Christian Scriptures, is an universal religion, which is destined to spread over the whole earth, and to embrace every created people in one fold under one shepherd—the substitution of Moses and Solomon for the two Saint Johns, is in fact producing and perpetuating the very evil which the alteration was professedly introduced to avoid—it is identifying the Order with *a peculiar religion*, which, though true at its original promulgation, was superseded by its divine author when the Sceptre had departed from Judah.

At this period the religious atmosphere was enlightened by THE BIRTH OF LIGHT, in the appearance of Shiloh—the Day Star from on high—a Light to lighten the Gentiles, and the glory of the people of Israel; who introduced a new Covenant, of which the religion of the Jews was a type or symbol; except that as the Mosaic dispensation was temporary, that of Christ was general, for all nations, and everlasting, for all ages.

LECTURE IV.

Epistle Dedicatory

TO

BRO.	JOSEPH EDWARDS CARPENTER,	W. M.
—	SAMUEL OLDHAM,	S. W.
—	JOHN BURGESS,	J. W.
—	SIR CHARLES DOUGLAS, M. P.	P. M.
—	CHARLES HARRIS,	TREA.
—	GEORGE J. KAIN,	SEC.
—	BREZZI ——,	S. D.
—	REV. W. WESTALL,	J. D.
—	JAMES SHARP, JUN.,	P. G. PURS.

Of the Shakespere Lodge Warwick, No. 356.

MY DEAR BRETHREN,

We live in strange, eventful times. Were our forefathers to rise from their graves, they would hold up their hands in astonishment, and pronounce it to be a different world from that which they had left behind them half a century ago. The work of locomotion, for which they were indebted to the power of living animals, is now effected by means of steam produced from a mineral dug out of the bowels of the earth; and even our artificial light, for which, at that period, a dead animal contributed various portions of its body, is also the result of a different combination of the same material.

The science of Chymistry has been the parent of both. And by the aid of another science, Electricity, we are enabled to hold familiar converse with friends at incredible distances, without any fear of interruption or disappointment by the miscarriage of letters, or the unfaithfulness or death of messengers. A taste for the fine arts is propagated throughout the whole population by Schools of Design; and a knowledge of general literature is diffused by means of itinerant lecturers, whose disquisitions are calculated to improve the reasoning faculty, and elevate the mind of man to its proper station, as a rational being created for immortality.

Amidst all this mass of moral and physical good, it behoves us, Free and Accepted Masons, to consider whether the Order we profess and admire is in a progressive state commensurate with the gigantic strides by which other sciences are advancing. The strife is antagonistic; and if we mean to gain the prize—if we have any ambition to win the applause of our contemporaries here, or to share in the rewards of successful diligence hereafter, we must forget those things that are behind, and reaching forth towards those things which are before, we must press forward towards the mark.

The Landmarks of masonry are necessarily stationary; for by a fundamental law of the Craft, they cannot be altered. To the Lectures, therefore, we must look for an evidence of the progressive improvement of the Order. And, accordingly, between the years 1717 and 1817, we have had six different arrangements of the Lectures, each being an improvement on its predecessor. But from 1814 to 1849, during which period such vast and momentous discoveries in science have been accomplished, our means of social improvement pursue the same unvaried round; and I am not cognizant that the Lectures have received a single alteration under the sanction of the Grand Lodge. If this apathy should unfortunately be of much longer duration, Freemasonry will soon be behind the times; and I am afraid its claim to a very remote origin will scarcely save it from neglect. A consummation which every true lover of the Order will most cordially deprecate.

The doctrines contained in the ordinary disquisitions of our lodges, I have endeavoured to embody in the following Lecture, which I trust you will accept as an offering of fraternal friendship and gratitude for kindnesses received; and by so doing you will confer an additional obligation on,

Dear Brethren,
Your faithful and obedient Servant,
GEO. OLIVER, D.D.,
Honorary Member of the Lodge.

SCOPWICK VICARAGE,
September 1, 1849.

Lecture the Fourth.

An examination of the Doctrines contained in the Lodge Lectures.

"When to the Lodge we go, that happy place,
There faithful Friendship smiles in every face.
What though our joys are hid from public view,
They on reflection please, and must be true.
The Lodge the social virtues fondly love;
There Wisdom's rules we trace, and so improve;
There we, in moral architecture skilled,
Dungeons for vice—for virtue temples build;
Whilst sceptered Reason from her steady throne
Well pleased surveys us all, and makes us one."

PROLOGUE AT EXETER, 1771.

"Rewarde the just, be steadfast, true, and plaine,
Represse the proud, maintaining aye the right;
Walke always so, as ever in his sight,
Who guardes the godly, plaguing the prophane.
And so ye shall in princely vertues shine
Resembling right your mightie King divine."

KING JAMES I.

THE Lectures of Masonry contain an extensive reference to a system of moral duties, applicable to every station of life, and to all situations in which a Mason can possibly be placed, although the explanations are not so ample and diffusive as might be wished. They apply in general to the duties which we owe to God, our neighbour, and ourselves; including brief dissertations on the theological and cardinal virtues; on Brotherly Love, Relief, and Truth; and a variety of other subjects connected with Bible history; for above all other means of promoting the interests of morality amongst the Brotherhood, it is felt that the influence of religion is the most efficient and certain; and hence the system is based on the knowledge and acknowledgment of a God who is the creator of the world, and the author and giver of every good and perfect gift.

Freemasonry must not, however, be mistaken for a religious sect, although it embraces that universal system in which all men agree; while the infidel and atheist are excluded, because they prefer the dangerous alternative of disbelieving the divine existence. It was on this rock that the efforts of Barruel, and some other opponents of Masonry, who, labouring to neutralize and destroy its influence, were miserably shipwrecked amongst the quicksands of ignorance and error. They endeavoured to make out that it was a religious sect of a most exclusive nature, founded on the principles of deism; and, failing in their proofs, all the elaborate superstructure which they had raised with so much labour and toil on this sandy foundation, fell to the ground in ruins.

The trust of a Mason is in the Most High God, as a basis which can never fail, and a rock which never can be shaken. Nor is it a mere empty profession; for it is borne out and illustrated by our practice. We open and close our Lodges with prayer; the same formula is used at the initiation of candidates; and no business of any importance is conducted without invoking the Divine assistance on our labours; and the blessing of God cannot be expected to follow any man's profession, unless it be verified by a good and virtuous life.

The Lectures of Freemasonry inculcate and enforce brotherly love as a means of inciting the performance of duty and the promotion of social happiness. But this doctrine, beautiful as it is, was thrown into shade by the introduction of the Theological Virtues, which display a Love that is divine; the operation of which, even the extinction of Faith and Hope, will not affect it in the slightest degree. This virtue is immortal, and will form the chief attraction amidst those everlasting hills which lie beyond the summit of the Masonic Ladder. T G A O T U himself taught it as it was never understood before; and by his instructions Masonry became an universal science; for he promulgated that great truth which Jew and Gentile alike refused to admit, that all mankind are brethren. No matter what may be the birth, language, or colour of the skin, every man is a brother if he faithfully performs his duty to God, his neighbour, and himself.

And what are all other connections when compared with this great principle? It is true, friendship is exceedingly pleasant; the relations of husband and wife, parent and child, are still more dear. But they fall into insignificance before the glorious immunity of being a brother to the whole human race; and by consequence, a son of God. And every one who keeps the commandments delivered to Moses on Mount Sinai, is entitled to participate in all the privileges of the Covenant made with the general Father of us all, by which we are constituted his children. A blessed fraternity; consecrated by Faith; supported by Hope; and cemented by universal Charity.

Now, in the arrangements of this world, a parent is generally desirous of having his children about him; and when arrived at years of maturity, he becomes solicitous to advance them to a prosperous situation in the world. And he rejoices when he finds them sober, industrious, and respected by worthy men. Is it not equally credible that our Father which is in heaven should rejoice when his children obey the truth; and that he is desirous of seeing them in the Grand Lodge above, standing round about his throne.

Again, all earthly connections, how endearing soever they may be, have their portion of sorrow. Friends may deceive us, and bring us into trouble. A beloved child may pursue evil courses, and bring down upon himself shame, and on his parents grief and care. A contentious wife or husband embitters life, and proves the uncertainty of perfect happiness on earth. And there is another consideration urged by Freemasonry, which may stimulate us to prepare assiduously for a state where pain, and care, and disappointment have no existence; and that is, the uncertainty which attends the enjoyment of all human felicity; symbolized in a Mason's Lodge by the legend of the third degree.

The relation of a father to a child, or of a husband to his wife, is delightful. But how excruciating is the sorrow with which the death of any one of them covers the survivors. Observe the weeping children at their father's grave;—observe the father, overwhelmed with grief, crying out bitterly, "My son, my son, would to God I had died for thee;"—observe a distracted mother, who will not be comforted, because her child is dead. Such sorrows attend all temporal pleasures and comforts; as is indicated by the Mosaic pavement of a lodge. We are happy to-day—we may be miserable to-morrow. If we are worldly, we are sure to feel the sorrows of the world. Is it not better to be spiritual, as all good Masons ought to be, who value their privileges and respect their obligation.

By such reasoning the doctrines of Freemasonry are brought to bear on the duties of social life; and if our sublime Order did not improve the condition of man as a citizen of the world, its practice would be an useless waste of time, and would scarcely deserve the attention of any conscientious brother.

In the lectures we are taught to consider faith in T G A O T U, which constitutes the first step of the Masonic Ladder, as primarily necessary to the attainment of its summit. But Faith must be shown by its fruits. It must lead through the portal of Hope to the consummation of Charity; for it is by the practice of morality that our faith is shown to be sincere, and made capable of producing a most refined Love. The duties taught in the Lectures of Freemasonry extend to almost every transaction of our lives, and contain rules which we may apply to every situation in which we are likely to be placed. Some of our opponents will affirm that morality is an useless qualification for any man to possess; and that nothing is required to prepare us for the happiness of another world, but a simple profession of faith. But if this be true, why did Jehovah promulgate the moral law with such ceremonial pomp? Did he reveal it with such tremendous accompaniments, without exacting obedience to its precepts? So far from it, that its obligations have been retained in the system of Christianity. The Redeemer of mankind commanded his followers to love the Lord their God with all their heart, and soul, and strength. This He pronounced to be the first great commandment. And the second is like unto it—thou shalt love thy neighbour as thyself.

For these and other reasons equally cogent, Freemasonry teaches her members that if God had not intended the commandments of the moral law to be eternally binding on the consciences of his creatures, he would not have revealed them with such circumstantial ceremony. Nor would the moral law of the Jews have been made the basis of our own system of religion, if Faith, the first step of the ladder, were capable, unaided, of giving us the victory. Other resplendent virtues must be superadded. And therefore the lectures of masonry enjoin that "as prudence directs us in this election of the means most proper to attain our ends, so justice teaches us to propose to ourselves such ends only as are consistent with our several relations to society, rendering to all without distinction those dues which they are respectively entitled to claim from us; bending with implicit obedience to the will of our Creator, and being scrupulously attentive to the sacred duties of life; zealous in our attachments to our native country; exemplary in our allegiance to the government under which we reside; treating our superiors with reverence, our equals with kindness, and to our inferiors extending the benefit of admonition, instruction, and protection."

As these and other similar doctrines and duties form a part of the very first Charge which is delivered to the incipient Freemason, it may be useful to take a brief view of them, as they apply to God, our neighbour, and ourselves; as they form an exemplification of the moral law which was communicated to our Grand Master Moses from the mountain where the Deity had previously manifested himself in a Burning Bush. The first Table of this law describes our duty to God; and the second our duty to our neighbour and ourselves. Let us, then, see how intimately these duties correspond with the teaching of our noble Order.

Freemasonry directs us to put our sole trust in the One God who dwelleth in the highest heavens, under the several names, in consecutive degrees, of Great Architect, Grand Geometrician of the Universe, and Most High or Jehovah. And teaches the true brother that "every blade of grass which covers the field, every flower which blows, and ever insect that wings its way in the bounds of expanded space, proves the existence of a First Cause, and yields pleasure to the intelligent mind. Were the enquiring mason to descend into the bowels of the earth, and explore the kingdom of ores, minerals, fossils, he would find the same instances of divine wisdom and goodness displayed in their formation and structure;—every gem and pebble proclaims the handywork of an Almighty Creator. Should he exalt his view to the more noble and elevated parts of nature, and survey the celestial orbs, how would his astonishment be increased. If, on the principles of Freemasonry and true philosophy, he contemplates the sun, the moon, the stars, and the whole concave vault of heaven, his pride will be humbled, and he will be lost in awful admiration. The immense magnitude of those bodies, the regularity and rapidity of their motions, and the vast extent of space through which they move, are equally inconceivable; and as far as they exceed human comprehension, baffle his most daring ambition, till, lost in the immensity of the theme, he sinks into his primitive insignificance. To him, the Great Geometrician of the Universe, the father of Light and Life, the fountain of eternal wisdom, let us humbly dedicate our labours; imploring him to bless and prosper the work of our hands, to his own glory, the good of mankind, and the salvation of our immortal souls."

Our Lectures teach this; and further add that we are not to defile ourselves by the worship of any other deity, but confine our adoration to Him alone. If it be asked how is it possible to worship any other gods? the answer is ready. By elevating gold into an idol, like the miser; who worships the glittering root of evil night and day; who thinks of nothing else, and hopes for nothing else.

Here, then, is an object which takes precedence of Jehovah. And I am afraid it is an idol which is worshipped by others besides the confirmed miser.

Again; how many are there who make an idol of pleasure; and even neglect the Most High, on his own sacred day, to follow it. In a word, it will be found a living fact, that whatever is inordinately loved and followed to the neglect of the allegiance which is due to the Divinity, is an idol and constitutes a breach of that preliminary law which the lectures of Freemasonry strictly enjoin the fraternity to keep unbroken.

It will not be difficult to ascertain whether such conduct be pursued by many of those who are strangers to our Order; and alas, by some who enjoy the privilege of its instructions. If, instead of imploring the aid of the

Great Architect of the Universe in all his undertakings, some reckless brother should prefer the things of this world;—if instead of looking up to Him in every emergency for comfort and support he should allow himself to be governed by the crude opinions or the interested recommendations of mankind; if instead of relying on Him in all doubts and difficulties, with the awe and reverence which is due from the creature to the Creator, he should unfortunately depend on his own understanding and trust to his own judgment;—if instead of making it his study to obey the divine will and pleasure, he should be inclined to consider his own interest or convenience in preference to it—we may be quite sure that the world is his idol, and business or pleasure, in his estimation, are preferable to the decrees of a wise and benevolent Creator.

Hence the propriety of the admonition to beware lest we bow down to a graven image, which is the second point noticed in the moral law, and constitutes a breach of masonic obligation. It may, indeed, be supposed improbable, in these enlightened times, that any one would be so weak and so absurd, as to fall down and worship any inanimate substance. The thing appears preposterous; but it is clear that the danger does not lie in the literal object of adoration. Whatever we covet inordinately, how trifling soever it may be, is an insult to the Deity. And therefore the Lectures of masonry teach us to subdue all our irregular passions and propensities, that a habit of virtue may be induced, to enlighten the mind and purify the soul.

It may therefore be reasonably concluded that whoever loves anything better than his duty to God, to him it is a graven image in which he puts his trust; however he may deceive himself and others at his initiation, by declaring that he puts his trust in God. One of the duties which we owe to that divine Being as taught by Freemasonry, is to hold his Sacred Name in the utmost reverence. And some of the superior degrees account it Ineffable, and not to be profaned by allowing it utterance, except on the most solemn occasions; as the high priest of the Jews was only allowed to enter the Sanctum Sanctorum on the great day of annual atonement; because the Almighty proclaims that "he will not hold him guiltless that taketh his Name in vain."

The Lectures further teach that our ancient brethren, after diligently attending to their worldly business six days in the week, devoted the seventh to rest and worship; and that it will be equally our duty and interest to copy their example; because the profanation of the Sabbath is forbidden in the Sacred Roll of the Law; and it unfortunately constitutes one of the crying sins of the nineteenth century. To give a better insight into the obligations of a Free and Accepted Mason on this point, I will suggest a few heads of self-enquiry to ascertain how far the above duties are binding on the fraternity; first producing a passage from a copy of the ancient Gothic Charges of Freemasonry already referred to. "Loveday (Sabbath) yet schul they make nonn, tyl that the werke day be clene a gonn, apon the holy day ye mowe wel take leyser y nowygh Loveday to make, lest that hyt wolde the werke day, latte here werke for suche a fray; to suche ende thenne that ye hem drawe, that they stonde wel yn Goddes lawe." And again. "Holy Churche ys Goddes hous, that ys y mad for nothynge ellus but for to pray yn, as the bok tellus; ther the pepul schal gedur ynne, to pray and wepe for here synne."

Do you, in obedience to this advice, conscientiously devote the Sabbath day to God, and spend its sacred hours in worship, in reading, and in meditation; or do you neglect its duties, and devote it to worldly or carnal purposes? The ancient masonic directions on this point, from the tenth to the fourteenth century, are plain and precise. "In Holy Churche lef nyse wordes of lewed speche and fowle wordes, and putte away alle vanyte, and say thy pater noster and thyne ave; loke also thou make no bere, but ay to be yn thy prayere, yef thou wolt not thyselve pray, latte non other no way. In that place nowther sytte ny stonde, but knele fayr down on the gronde, and when the Gospel me rede schal, fayre thou stonde up fro the wal, and blesse the fayre, yef that thou conne, when Gloria tibi is begonne; and when the Gospel ys y doun, agayn thou mygth knele adown—on bothe thy knen down thou falle, *for hyse love that bought us alle.*"

It will, then, appear perfectly clear, that according to the teaching of ancient masonry, every abuse of that day which has been set apart as a day of rest, will be brought to a strict account; for if Jehovah has solemnly warned us to "remember that we keep holy the Sabbath day," he will not fail to remember also whether we have obeyed or disobeyed the command. What, indeed, can be a more glorious sight for men and angels, than an assembly of human creatures, who have souls to save, united in the public worship of God, as directed in the above old masonic manuscript.

Our Saviour laid this practical demand on the brethren; "a new commandment I give unto you, that ye love one another." And he adds: "by this shall all men know that ye are my disciples." This is the principal object of Freemasonry. It is, therefore, our indispensable duty to love the brethren; or in other words, to do each other all the good offices in our power; to be kind, compassionate, and charitable; not to speak evil of others, nor to listen when others wish to speak evil of their neighbours. It is, indeed, too true that there are many in the world who appear to take delight in promoting disputes, and fomenting quarrels. But is it agreeable to the rules of masonry, to set friends at variance with each other, to disunite families, and to throw society into confusion? On the contrary, its Lectures speak highly of the value of a tongue of good report; and recommend the fraternity to speak as well of a brother in his absence as in his presence; and, if unable to do so with a safe conscience, to preserve a strict and charitable silence, as the distinguishing virtue of the Order. Let the master of a Lodge, therefore, take every opportunity of cautioning the brethren not only to refrain from slander themselves, but never to listen to it; for it is a breach of the stringent injunction of the grand patron of masonry, St. John the Evangelist, who, in imitation of his divine Master, frequently directs us to love one another. And for this reason, because if the love of God and man be the ground of our actions, it will promote our own happiness, and spread the blessings of peace and unanimity amongst all ranks and descriptions of men. This is the spirit which the Lectures of Freemasonry inculcate. It is the disposition of angels and the practice of every faithful brother.

One of the numerous virtues which are strongly recommended in the masonic system, is filial piety. According to the ancient masonic record which has been so copiously exemplified in this Lecture, the reward of duteous children is thus stated. "These lordys chyldryn therto dede falle to lurne of hym the craft of Gemetry, which was the name of Masonry, the wheche he made ful curysly; throygh fadrys prayers and modrys also, thys onest craft he putte hem to; he that lerned best and were of oneste and passud hys felows yn curyste, gef yn that craft he dede hym passe, he schulde have more worschepe then the lasse."

The heathen knew very little of this virtue; for the custom of exposing infirm children

——————— voteque sæpe
Ad spurcos decepta lacus,

was not a recommendation to the brothers and sisters of the unfortunate victims to be kind to their parents in return; although the law of Solon, called Alimenta, made it incumbent on children to provide for their aged parents.

Amongst the later Jews this principle was virtually renounced; although they were commanded to "honour their father and mother, that it might be well with them, and that they might live long upon the earth." And the duty was frequently reiterated in their sacred writings. Thus the son Sirach says very affectingly, "honour thy father with thy whole heart, and forget not the sorrows of thy mother. Remember that thou wast begotten of them, and how canst thou recompense them the things which they have done for thee?" Their disregard to parental necessities, however, was so remarkably prominent as to be formally recognized by a Jewish statute, denominated Corban; and hence our Saviour reproaches the Jews for making a corban of that which ought to be appropriated to the use of their parents. Thus if a parent was in necessity, and applied to his child for assistance, the answer was, I have already devoted what you require of me to God, and therefore if I give it you, I shall be guilty of the greatest profanation. The Talmud gives the form of appropriation; and though it is contrary to reason and the feelings of nature, yet it was universally approved and practised by the Pharisees and their successors.

Amongst masons the great principle of filial piety is denoted by the word Lewis, which signifies strength; and in operative masonry consists of certain pieces of metal, which, when dovetailed into a stone, form a cramp by which great weights, otherwise immovable, are raised to certain heights with very little difficulty, and fixed on their proper bases. In speculative masonry, the word Lewis symbolizes the son of a master mason, whose duty it is to bear the burden and heat of the day in lieu of his parents, who, by reason of their age, ought to be exempt; to help them in the time of need, and thereby render the close of their days happy and comfortable. His privilege for so doing is to be made a mason before any other person, however dignified.

These precepts and duties are incumbent upon all good and dutiful children who have been initiated into the mysteries of masonry. They must honour their parents by a filial affection for their persons, a deference to their opinions, a tender regard for their safety, and an implicit obedience to their commands. Being fully sensible of the immense obligations which their nurture and education have imposed upon them, they feel themselves at all times ready to show their respect by a constant and cheerful attendance to their wishes. If their parents are in want, they relieve them; if they are feeble or infirm, they support them, and on every occasion are prepared to vindicate their welfare and happiness.

Such are the obligations which masonry imposes, and they are well calculated to contribute to the general good of society; for they are founded upon the best feelings of our nature, and not only insure domestic happiness, but also harmony and peace amongst all ranks and descriptions of men. A parent has claims upon his child which none but a parent can know. No other can estimate the affectionate care and anxiety which dwell in the heart of a parent to produce the welfare of his offspring, at a time when they are in capable of providing for themselves—when they must have perished, if deprived of parental tenderness. The wise man truly says, "the father waketh for the daughter when no man knoweth, and the care for her taketh away sleep." Common gratitude, therefore, ought to point out the necessity of an adequate return, when age and infirmity have cast a shade over these early and constant friends, and the loss of health and strength is accompanied by adversity, and perhaps by sickness and pain.

But experience shows that gratitude is too slender a tie to operate with a beneficial effect on stubborn or impracticable natures. We frequently see parents deserted by their children at a period when they are most in need of active assistance. They have arrived at maturity perhaps—they have families of their own to provide for—and this is considered a sufficient excuse for leaving their aged parents to the mercy of the world. They have been indebted to them for nurture, for instruction, for the means of procuring subsistence; and yet all these benefits are overlooked on a narrow and selfish principle which Freemasonry is solicitous to remove. They abandon their parents to want and all its miserable attendants; and by such conduct, frequently bring their gray hairs with sorrow to the grave.

The Mosaic pavement of a mason's lodge is placed there to show the vicissitudes of human life; that however prosperity may favour us with its smiles to-day, it is uncertain how long it will continue to bless us. Adversity may come when we least expect it, and penury and distress may follow joy and pleasure. The latter period of life may be subjected to want and misery, when we are most unfit to encounter it; and instead of resting in peace after a long and troublesome journey, we may be compelled again to encounter the burden and heat of the day. This, then, is the period for the Lewis to display the virtues of filial piety and gratitude; and in such a case no danger is so great, but he will readily encounter it, and no toil so severe, but he will willingly bear it. This is strongly recommended in the system of Freemasonry; and forms an important link in the chain of benefits which society receives from this benevolent institution.

Now, as Free and Accepted Masons, we may be inclined to believe that we have done every thing which the Order prescribes in this particular, by obedience and submission to our immediate parents. But it is our duty also to honour all men, and love the brotherhood, whatever be their station, whether superiors, equals, or inferiors. To the first we owe submission; to the two last love and condescension. It will be for us to consider whether we have always obeyed this command, by submitting to lawful authority on the one hand, or treating our humbler brethren with supercilious contempt on the other. What alas! is the greatness of this world? All that can be said of the very proudest man in existence, after he is dead, is—that he was born—he lived—and he died. Pride is as hateful to God as it is to man. He made the poor as well as the rich; and with him there is no respect of persons.

The doctrines of masonry respect equally a brother's life; the chastity of his wife and daughter; and the protection of his property and reputation. A brother's life is a sacred deposit, which no one will dare to violate, if he duly reflects on the punishment of murder, as it is delineated in certain ceremonies which are familiar to the perfectly initiated mason. But there are some things more valuable than life. The Book which lies open on the pedestal of a mason's lodge will tell you what they are. "Ye have heard that it was said by them of old time, thou shalt not kill; and whosoever shall kill shall be in danger of the judgment. But I say unto you, that whosoever is angry with his Brother without a cause, shall be in danger of the judgment; and whosoever shall say to his Brother, Raca, shall be in danger of the council; but whosoever shall say, thou fool, shall be in danger of hell fire." By these words we understand that anger and quarrelling are highly reprehensible, and therefore our laws contain stringent provisions against all such improper practices. Our Grand Patron St. John is particularly strong upon this point, when he says "he that loveth not his brother abideth in death." And he has a plainer expression even than this, and more to our present purpose. "Whosoever hateth his brother is a murderer; and no murderer hath eternal life abiding in him."

Who, then, can be innocent of this offence? Where is the man to be found who has never been at variance with his brother? And this, on the above authority, is called murder. And, indeed, hatred and illwill have often ended in murder, when the angry man deemed himself incapable of such a crime. Consider the delinquency of the three Fellow Crafts, and reflect on their punishment. And if we suffer ourselves to be provoked to anger on every trifling occasion, who can tell what injury may arise, if our anger be increased by insult and provocation.

But there are many other ways of violating this precept, against which the Free and Accepted Mason ought to be for ever on his guard. For instance; in wishing for the death of any person who may stand in the way of our advancement; or that of any official person whom we hope to succeed. This is at least a violation of our duty to God, our neighbour, and ourselves; and a breach of our masonic obligation.

The next precept to which I have referred, as being one of the peculiar doctrines which are contained in the lectures of masonry, respects the personal chastity of a mason's wife or daughter. Indeed it was a positive command written by the finger of God—"Thou shalt not commit adultery." And a primitive law of masonry was thus expressed.

Thou schal not by thy maystres wyf ly,
Ny by thy felows, yn no maner wyse,
Lest the craft wolde the despyse;
Ny by thy felows concubyne,
No more thou woldest he dede by thyne.
Gef he forfete yn eny of hem,
So y chasted thenne most he ben;
Ful mekell care mygth ther begynne,
For suche a fowle dedely synne.

There is every reason to believe that this law is respected by the fraternity, although it condemns a sin which is sometimes practised by others without remorse, and talked of without shame. Our Grand Master, king Solomon, however, says, "rejoice O young man, and let thy heart cheer thee in the days of thy youth, and walk in the ways of thine heart and in the sight of thine eyes;

but know thou that for all these things God will bring thee into judgment."

We now come to the consideration of the inviolability of our neighbour's property and reputation, which Freemasonry binds us to protect as carefully as if they were our own. It is not enough that no encroachments are made on it by actual violence on our own part, or connivance when it is invaded by others. The good mason, who acts according to the instruction which he receives in the Lodge, will not endeavour to profit by his brother's ignorance or inexperience in any worldly transactions; but so far from injuring him, he will protect him from danger by giving him timely notice of any attack which may be meditated by others; thus showing to the uninitiated world that we are united by a chain of indissoluble affection, which cannot fail to distinguish us while we continue to practise the distinguishing duties of our profession, Brotherly Love, Relief, and Truth.

In the United States, there is, or was, a peculiar degree, by which the fraternity undertook to protect the interests of their brethren from encroachment; and for this purpose, the candidate, at his admission, promises to caution his brother by sign, word, or token, not only when he is about to do anything contrary to the principles of masonry, or whenever he sees him about to injure himself by inadvertence or ignorance in buying and selling; but that he will himself, being so cautioned, pause and reflect on the course he is pursuing; and that he will assist a brother by introducing him in business to his friends, and to promote his interests by every means in his power.

And further, Freemasonry teaches the brethren by its lectures that it is their duty to support a brother's character in his absence more energetically than in his presence, because it is presumed that when present he will be able to vindicate his own reputation; not wrongfully to revile him, or suffer him to be reviled by others if it is in their power to prevent it. An offence against this precept is committed by misrepresenting the conduct of a brother, or passing uncharitable reflections upon him; by whispering, backbiting or circulating injurious reports. This has always been considered a fault of such magnitude, that T G A O T U has decreed that "all liars shall have their part in the lake which burneth with fire and brimstone;" and Freemasonry discountenances slander by giving honour to the tongue of good report, which is indicated in the Tracing Board by a specific and expressive symbol.

I shall now consider a few negative virtues which are embodied in the lectures of Freemasonry; trusting that they will be found equally valuable with those already mentioned; and applicable alike to the fraternity, and to society at large; because an abstinence from vice is as useful to the community as the practice of virtue.

The best masons are not always the most learned men; for the chief excellence of the Order does not consist so much in its science as in its morality. The moral and intellectual, though frequently found in unison with each other, are qualities essentially different. "That virtue proceeds from rectitude, and vice from error of judgment, we do not at all, perhaps, perceive with sufficient clearness. By the terms of distinction we are in the habit of using in familiar discourse, when we are speaking of the intellectual and moral characters of mankind, we are some of us, possibly, in some measure diverted from discerning the derivation of right and wrong conduct, from just and false opinion. We speak, in common conversation, of a good head and a good heart; and we are carelessly led by this local account of intellectual and moral excellence, to conceive of good sense and good living, as proceeding from different departments and provinces of our nature; and fancifully to consider them as having their source where we thus figuratively assign them their seat."[1]

But universal experience convinces us that this is not always the case. The cleverest and most talented men are sometimes extremely dissolute. Such persons, when they apply their learning or talent to improper purposes, are the most dangerous characters in existence. Their judgment and tact enable them to conceal the iniquity of their intentions, and they hence practise their nefarious plans on the credulity of the public with impunity. Joseph Balsamo, better known under his assumed name of Count Cagliostro, was a clever charlatan of

[1] Fawcett's Sermons, vol. i., p. 127.

this description. His success in the prosecution of his schemes is thus detailed in the memoirs of Abbé Georgel, touching the case of Cardinal Rohan. "In the mean time an unfortunate circumstance contributed to hurry the Cardinal into extraordinary adventures. I do not know what monster, envious of the tranquility of honest men, had vomited forth upon our country *an enthusiastic empiric,—a new Apostle of the religion of nature, who created converts in the most despotic manner, and subjected them entirely to his influence.*

"Some speedy cures effected in cases that were pronounced incurable, and fatal in Switzerland and Strasburg, spread the name of Cagliostro far and wide, and raised his renown to that of a truly miraculous physician. His attention towards the poor and his contempt for the rich, gave his character an air of superiority and interest which excited the greatest enthusiasm. Those whom he chose to honour with his familiarity, left his society with ecstacy at his transcendent qualities. The Cardinal de Rohan waa at his residence at Saverne, when the Count de Cagliostro astonished Strasburg and all Switzerland with the extraordinary cures he performed. Curious to see so remarkable a personage, the Cardinal went to Strasburg. It was found necessary to use interest to be admitted to the Count. If M. le Cardinal is sick, said he, let him come to me and I will cure him; if he be well, he has no business with me, nor have I with him. This reply, far from giving offence to the vanity of the Cardinal, only increased the desire he had to be acquainted with him.

"At length, having gained admission to the sanctuary of this new Esculapius, he saw on the countenance of this incommunicative man a dignity so imposing that he felt himself penetrated by a religious awe, and that his first words were inspired by reverence. This interview, which was very short, excited more strongly than ever the desire of a more intimate acquaintance. At length it was obtained, and the crafty empiric timed his conduct and his advances so well, that at length, without seeming to desire it, he gained the entire confidence of the Cardinal, and possessed the greatest ascendency over him. His Egyptian lodges were opened at night in the Cardinal's own drawing room, illuminated by an immense number of wax tapers; and he succeeded in persuading his dupe, that under the influence of a familiar demon, he could teach him to make gold out of baser metals, and transmute small diamonds into large precious stones. And thus under the pretence of developing the rarest secrets of the Rosicrucians and other visionaries, who believed in the existence of the Philosopher's stone, the elixir of life, &c., he cheated the Cardinal out of large sums of money, which, instead of passing through the crucibles, found their way into the pockets of the sharper."

The true science of Freemasonry guards against such impostures by the most stringent regulations; and recommends the practice of virtue as a shield against the impositions of designing men. Thus the Constitutions declare that no person shall be admitted as a candidate without notice and strict enquiry into his character and qualifications. That every candidate must be a free man and his own master, and at the time of his initiation, be known to be in reputable circumstances. He should be a lover of the liberal arts and sciences, and have made some progress in one or other of them. And previous to his initiation, he is called upon to subscribe a declaration that he will cheerfully conform to all the ancient usages and established customs of the Order. And even then, he cannot on any pretence be admitted, if, on the ballot, three black balls shall appear against him.

The Charge delivered to an entered apprentice is equally plain and significant. "No institution can boast a more solid foundation than that on which Freemasonry rests—*the practice of social and moral virtue;* and to so high an eminence has its credit been advanced, that in every age, monarchs themselves have become the promoters of the art, have not thought it derogatory from their dignity to exchange the sceptre for the trowel; have patronized our mysteries, and even joined our assemblies."

With such precautions, it will appear at least very improbable that vicious or unworthy characters should gain admission into a lodge. But unfortunately this does sometimes occur. And the characters of men undergo such extraordinary and unexpected changes by the force of circumstances, that it is impossible for Freemasonry to answer for the stability of every brother who may have been enrolled amongst its members; and such alterations in the disposition cannot be provided against by any sumptuary law. The universal system remains unsullied by the introduction of an occasional impostor, although the locality where such an event occurs may suffer a temporary shadow to obscure its light; for as Agesilaus observed when the director of ceremonies in the Gymnasium placed him in an unworthy situation; "it is not the place that makes the man, but the man that makes the place honourable or dishonourable."

It is to be regretted, however, that such impostors sully and avert the stream of masonic charity. They prowl about the country with false certificates, and often succeed in deluding benevolent brothers, to the injury of those who are really in distress. And the difficulty of distinguishing between real and assumed objects of charity may be estimated from such cases as the following, which has been extracted from the Quarterly Communication for Dec., 1823. "A Report from the Board of General Purposes was read, stating that an individual, calling himself Simon Ramus, had been endeavouring to impose upon the brethren, and to obtain pecuniary assistance, under colour of a fabricated certificate, stating him to have been a member of the Lodge No. 353. And also that another individual, calling himself Miles Martin, but supposed to be one Joseph Larkins, had, in a similar manner, been endeavoring to impose upon the brethren, under colour of a certificate from the Grand Lodge of Ireland and another from the Lodge No. 145, at Norwich; all which certificates had been detained and transmitted to the Grand Lodge. The Board stated that they were induced to make this Report with a view to guard brethren against further attempts at imposition by those individuals, although their means were in a great measure destroyed by the detention of the certificates."

Such cases are of common occurrence in the country; and to guard against them as completely as possible, the laws, under the presumption that ignorance is the parent of vice, provide against the admission of uneducated persons who are incapable of writing their own names, by requiring them actually to suscribe the Declaration. A want of attention to this rule is calculated to produce many other irregularities. The following censure of the Grand Lodge on this point merits general circulation amongst the brethren. For obvious reasons the name of the offending lodge is omitted. "It being remarked in the Grand Lodge that some of the brethren of the Lodge No. — were unable to write, inasmuch as their marks only were affixed against their names, and amongst them was the Junior Warden; and the law, sec. iv., p. 90, declaring such individuals ineligible for initiation, the M. W. Grand Master will, after this notice, feel it a duty he owes to the Craft to bring under the cognizance of the Grand Lodge the conduct of any Lodge which shall violate the wholesome and necessary law above referred to; a breach of which it is declared in the preamble to the regulations for proposing members, &c., p. 88, shall subject the offending lodge to erasure. And the M. W. Grand Master will require his Provincial Grand Masters to warn the lodges under their respective superintendence, of this His Royal Highness's determination, and to report to him any instance which shall come to their knowledge of a disregard of the law in this respect."[2]

[2] Quart. Com. 26th Sept., 1826.

LECTURE V.

Epistle Dedicatory

TO

BRO.	REV. CHARLES NAIRNE, D. P. G. M.	& W. M.
—	REV. W. N. JEPSON,	S. W.
—	G. T. W. SIBTHORP, ESQ.,	J. W.
—	E. F. BROADBENT, ESQ.,	P. M.
—	G. H. SHIPLEY,	TREA.
—	RALPH TAYLOR,	P M. & SEC.
—	M. WOODCOCK,	S. D.
—	JOSEPH DURANCE,	J. D.
—	JOHN MIDDLETON,	P. M. & M. C.
—	HENRY COTTON,	STEWARD,

Of the Witham Lodge, Lincoln, No. 374.

MY DEAR BRETHREN AND FRIENDS,

Whom I know so well, and esteem so highly, will accept this trifling testimony of my regard, resulting from a connection of many years' standing, and a social intercourse that has, I flatter myself, been mutually advantageous.

Oft have I met your social band,
  And spent the cheerful festive night;
Oft, honoured with supreme command,
  Presided o'er the sons of Light.

And by that hieroglyphic bright
  Which none but craftsmen ever saw;
Strong memory on my heart shall write
  Those happy scenes when far awa'.

BURNS.

Those happy days are gone, never to return. Younger men may enjoy them as I have done; although it will require a very high degree of enthusiasm to impart the same relish which it has been my good fortune to possess. Some of my happiest moments have been passed in a mason's lodge. Commonplace maxims, if they were only clothed in the mantle of Freemasonry, have been invested with all the dignity of philosophy; and prosaic precepts have mounted in my excited imagination to the sublime regions of poetic inspiration.

I often look back upon that period with sensations of unfeigned pleasure. It was like a bright halo of glory which overshadowed my path, and cast its streams of glittering light about me. The Lodge was a Paradise of pleasure, and masonry spread a gleam of sunshine on my existence. Your lodge, my dear brethren, is associated in my mind with that season of felicity, as connected with my Provincial presidency; and I cannot offer you a better wish than that you may enjoy, as I have done, the social delights which spring from an intercourse with each other in a tyled Lodge, governed by Wisdom, protected by Strength, and enlivened by Beauty.

There are many good masons in the Witham Lodge who have made it their study to investigate the doctrines contained in the system of Freemasonry, and to trace them to their source, that their accuracy may be established by the most unexceptionable references. If it were not founded on the strong basis of truth, it would not have so nobly and triumphantly sustained the repeated attempts, both direct and indirect, and all insidious, to sully its fair fame, and blot it out of the list of those beneficent institutions which confer so much honour on their supporters, and such inestimable benefits on society at large.

The symbol which forms the subject of the Lecture which I have the pleasure of dedicating to you, is of such importance as to admit of several interpretations, each illustrating the sublimity of its reference; and unitedly forming a constellation of moral and religious virtues which constitute an appropriate introduction to the divine qualities that point the way to heaven.

That the Members of the Witham Lodge may experience no difficulty in finding that narrow path, is the sincere wish of

W. Sir, and dear brethren,
Your sincere friend,
And faithful Brother,
GEO. OLIVER, D.D.,
Hon. Member of the Witham Lodge.

SCOPWICK VICARAGE,
October 1, 1849.

Lecture the Fifth.

The twelve definitions of the Circle and Parallel Lines considered.

"The Circle has ever been considered symbolic of the Deity; for as a Circle appears to have neither beginning nor end, it may justly be considered a type of the Deity, without either beginning of days or ending of years. It also reminds us of a future state, where we hope to enjoy everlasting happiness and glory."

OLD LECTURES

WHEN a candidate for masonic honours has been enlightened by describing the circle of duty round the central point of light, and is permitted to look round him and observe the appointments and decorations, he is particularly struck with the appearance of numerous instruments of mechanical labour which appertain to the trade of an operative mason; intermixed with a profuse sprinkling of astronomical signs, and indications of a present Deity, which he sees arranged methodically about the Lodge. On whatever side he may turn his eyes, the effect is still the same. The ceiling is covered with symbols; as well as the pedestal, the tables, the walls, and the pillars. And before him, on what he may take to be an altar, lies the Volume of the Sacred Law of God, covered with mathematical instruments; which he also observes are repeated on the bosoms of the Officers. These we call Jewels, for the same reason as righteous men are often dignified with this title in the Holy Volume just mentioned,[1] because they are held amongst us in superior estimation. The floor is composed of Mosaic work, and surrounded with a tesselated border, in imitation of the lithostrata or tesselated pavements of the

[1] Mal. iii., 17.

Romans, so many remains of which exist in our own country.

On a first view of this peculiarity, the ideas of the candidate may probably revert to the chambers of imagery, so well described by the prophet Ezekiel;[2] to the Pantheon of Rome, the Catacombs of Egypt, or the cavern temples of Hindoostan. But his antiquarian reminiscences will afford a very imperfect notion of the scene which is exhibited in a mason's Lodge; for those decorations, unlike the symbols of masonry, consisted of sensible objects of veneration—single and double-faced deities, compound and imaginary animals;

> Genii with heads of birds, hawks, ibis, drakes,
> Of lions, foxes, cats, fish, frogs, and snakes,
> Bulls, rams, and monkeys, hippopotami,
> With knife in paw, suspended from the sky;
> Gods germinating men, and men turn'd to gods,
> Seated in honour with gilt crooks and rods;
> Vast scarabæi, globes by hands upheld
> From chaos springing, 'mid an endless field
> Of forms grotesque—the sphynx, the crocodile,
> And other reptiles from the slime of Nile.[3]

Similar monuments have been recently discovered in Central America, which Stephens thus describes. They "stand in the depths of a tropical forest, silent and solemn, strange in design, excellent in sculpture, rich in ornament, different from the works of any other people, their uses and purposes, their whole history so entirely unknown, *with hieroglyphics explaining all*, but perfectly unintelligible. Often the imagination was pained in gazing at them. The tone which pervades the ruins is that of deep solemnity. An imaginative mind might be infected with superstitious feelings. From constantly calling them by that name in our intercourse with the Indians, we regarded those solemn memorials as *Idols*, deified kings and heroes; objects of adoration and ceremonial worship. We did not find on either of the monuments or sculptured fragments any delineations of human, or in fact, any other kind of sacrifice; but had no doubt that the large sculptured stone invariably found before each idol, was employed as a sacrificial altar

[2] Ezek. viii., 10. [3] Hall's Life of Salt, vol. ii., p. 416.

The form of sculpture most frequently met with was a death's head; sometimes the principal ornament, and sometimes only accessory; whole rows of them on the outer wall, adding gloom to the mystery of the place, keeping before the eyes of the living, death and the grave; and presenting the idea of a holy city—the Mecca or Jerusalem of an unknown people."[4]

And a most curious circumstance respecting these ancient monuments of a nation whose very existence is unrecorded, is worthy of notice here. The workmen in their construction, used certain mason-marks to indicate their own productions. Thus Stephens tells us that "on the walls of these desolate edifices were prints of the *mano colorado*, or red hand. Often as I saw this print, it never failed to interest me. It was the stamp of the living hand; and always brought me nearer to the builders of these cities. The Indians said it was the hand of the Master of the Building."[5]

In a mason's lodge, however, every thing which the candidate sees before his eyes, possesses a symbolical meaning to recommend the practice of virtue in order to produce the glory of God, peace on earth, and good will towards men; a result which is considered acceptable to T G A O T U, because it cannot fail to prove a source of happiness to his creatures, and lead to an abundant reward in the mansions of the blessed.

The explanations of one series of these masonic symbols will be amply sufficient to illustrate my proposition, that the poetry of Freemasonry should be understood and felt before the science can be estimated according to its real value; for symbolism constitutes not only the materials, but the very essence of poetry. The early Christians had a symbol for every thing; nor did they hesitate to borrow these expressive tokens from the heathen, if they promised to advance the interests of their own system of religion. As Lord Lindsay observes respecting the adoption of pagan rites and ceremonies into Christianity, "our ancestors touched nothing that they did not Christianize; they consecrated this visible world into a temple of God, of which the heavens were the dome, the mountains the altars, the forests the pil-

[4] Yucatan, vol. i., p. 158. [5] Ibid. vol. ii., p. 46.

lared aisles, the breath of spring the incense, and the running streams the music,—while in every tree they sheltered under, in every flower they looked down upon and loved, they recognized a virtue or a spell, a token of Christ's love to man, or a memorial of his martyr's sufferings. God was emphatically in all their thoughts, and from such, whatever might be their errors, God could not be far distant. It would be well for us could we retain that early freshness in association with a purer and more chastened creed."[6]

The symbols of masonry possess the quality, above those of any other society, of exalting, by the sublimity of their nature, and the aptitude of their application, the character of the Order to its highest point of beauty and usefulness. In a word, Freemasonry differs essentially from all other human societies, in its moral organization and benevolent character. This difference, as is well expressed by our transatlantic brother, the Rev. Salem Town, in his first Prize Address, is clearly marked, and may be distinctly perceived to lie, primarily, in the simplicity and obvious purity of its first principles, and subordinately, in their natural and perfect adaptedness to the end in view. A defect in either case would mar the whole, endanger the unity, and defeat the design. Sound principles, injudiciously or wrongfully applied, may utterly fail to accomplish a desirable and proper object, simply for the want of adaptedness in the means, to secure the end. Freemasonry is a well adjusted course of means, most wisely carried out by its members, in the accomplishment of specific acts of a benevolent nature.

The symbols which I have selected for my present purpose, are the Point within a Circle flanked by two perpendicular parallel lines, supporting the Holy Bible, on which rests the foot of a Ladder, containing staves or rounds innumerable; and three gates, with the figures of Faith, Hope, and Charity, at equal distances from each other on the ascent. Its summit penetrates the highest heavens, symbolized by a semicircle or Rainbow edged with the three prismatic colours; surmounted by the vesica piscis, and divine triangle containing the Sacred Name; while the host of heaven are represented

[6] Christian Art, vol. i., xxvii.

as singing "Glory to God in the highest, and on earth peace, good will towards men." In the firmament is seen a Blazing Star, and the Sun, Moon, and seven Planets of the ancient world.

This combination of symbols embraces the general plan and design of the masonic Order; which is, to teach the brethren so to use things temporal that they finally lose not the things that are eternal; or in other words, to instruct them how they may conduct themselves in their passage through this short and transitory life so as to entertain a just and reasonable hope at its conclusion, of receiving the joyful sentence of approval from the lips of a just but lenient Judge, "Come ye blessed of my Father, inherit the kingdom prepared for you from the beginning of the world."

To explain this glorious Symbol seriatim, we will first consider the situation where it is placed. Our ancient brethren, who reduced the scattered elements of Freemasonry into order at the beginning of the last century, considered the lodge to be situated in the valley of Jehoshaphat; and that in whatever part of the world it might be opened, it was still esteemed, in a figure, to occupy that celebrated locality. Thus it was pronounced in the earliest known lectures, that "the lodge stands upon holy ground, or the highest hill or lowest vale, or *in the Vale of Jehoshaphat.*" This celebrated valley derived its name from Jehovah and Shaphat, which mean *Christ* and *to judge;* and as the prophet Joel had predicted that the Lord would gather together all nations, and bring them down into the valley of Jehoshaphat, it was believed by the Jews, (and the Christians subsequently adopted the same opinion,) that in this place the transactions of the great day of Judgment would be enacted.

Thus in the ninth century, Bernard the Wise, a Christian pilgrim, in his Travels in the Holy Land, says, "in the valley of Jehoshaphat there is a Church of St. Leon, in which it is said that our Lord will come to the last judgment." Sir John Maundeville, speaking of the transfiguration on Mount Tabor, gives the tradition of his time (A.D. 1322) as follows: "On that hill (Mount Tabor) and in that same place, at doomsday, four angels shall blow with four trumpets, and raise all men, that have suffered death since the world was created, to life;

and they shall come in body and soul in judgment, before the face of our Lord, in the valley of Jehoshaphat. And it shall be on Easter day, the time of our Lord's resurrection; and the judgment shall begin on the same hour that our Lord descended to hell and despoiled it; for at that hour shall he despoil the world, and lead his chosen to bliss."[7] The Mahometans entertained a similar belief. "Upon the edge of the hill," says Maundrell, "on the

[7] His description of this valley is very curious and interesting. "In the middle of the Valley is a little river, which is called the brook Cedron; and across it lies a tree, of which the Cross of Christ was made, on which men passed over; and fast by it is a little pit in the earth, where the foot of the pillar still remains at which our Lord was first scourged; for he was scourged and shamefully treated in many places. Also in the middle of the valley of Jehoshaphat is the church of our Lady, which is forty-three steps below the sepulchre of our Lady, who was seventy-two years of age when she died. Beside this sepulchre is an altar where our Lord forgave St. Peter all his sins. From thence, toward the west, under an altar, is a well which comes out of the river of Paradise. You must know that that church is very low in the earth, and a part is quite within the earth. But I imagine that it was not founded so; but since Jerusalem has often been destroyed, and the wall beaten down and tumbled into the valley, and that they had been so filled again, and the ground raised, for that reason the church is so low within the earth. Nevertheless, men say there commonly, that the earth hath so been cloven since the time that our Lady was buried there; and men also say there, that it grows and increases every day, without doubt. Beside that church is a chapel, beside the rock called Gethsemane, where our Lord was kissed by Judas, and where he was taken by the Jews; and there our Lord left his disciples when he went to pray before his passion, when he prayed and said, O, my Father, if it be possible, let this cup pass from me. And when he came again to his disciples, he found them sleeping. And in that rock within the chapel we still see the marks of the fingers of our Lord's hand, when he put them on the rock when the Jews would have taken him. And a stone's cast from thence, to the south, is another chapel, where our Lord sweat drops of blood, and close to it is the tomb of King Jehoshaphat, from whom the valley takes its name. This Jehoshaphat was king of that country, and was converted by a hermit, who was a worthy man and did much good. A bow shot from thence to the south, is the church where St. James and Zachariah the prophet were buried. Above the vale is Mount Olivet, so called for the abundance of olives that grow there. That mount is higher than the city of Jerusalem; and therefore from that mount we may see many streets of the city. Between that mount and the city is only the valley of Jehoshaphat, which is not wide. From that mount our Lord Jesus Christ ascended to heaven on ascension-day, and yet there appears the imprint of his left foot in the stone. Below is the stone on which our Lord often sat when he preached; and upon that same shall he sit at the day of doom."

opposite side of the valley of Jehoshaphat, there runs along, in a direct line, the wall of the city, near the corner of which there is a short end of a pillar jutting out of the wall. Upon this pillar, the Turks have a tradition that Mohammed shall sit in judgment at the last day; and that all the world shall be gathered together in the valley below, to receive their doom from his mouth."

In this place, therefore, the people beheld in imagination the Throne of Glory amidst clouds and darkness, surrounded by angels and archangels and the host of heaven. Here they heard the trumpet sound piercing the depths of the earth and sea, and calling up the dead from the most hidden recesses of both, to be rewarded or punished according to their works. And hence the valley became the burial place of those favoured few who could procure the great privilege of interment in such a sacred spot.

The valley is now for the most part a rocky flat, with a few patches of earth here and there. The western side is formed by the high chalk cliff supporting the city wall, and the opposite side by the declivities of the Mounts of Olives and Offence. It was evidently a burying place of the ancient Jews from the number of old sepulchral remains and excavations which it offers. and which the Jews have neither the means nor power to execute since their own desolation. That it was the cemetery of their fathers, and that they here expect the final judgment to take place, is a sufficient inducement to desire to lay their bones in this valley. For this reason many of the more devout Hebrews resort to Jerusalem from all parts of the world, to die there, and to be buried in the valley of Jehoshaphat. For the privilege of interment in this venerated spot, immense prices are often paid to the exacting Turks, and not seldom a grave is stolen in the solitude and darkness of the night. The modern Jews content themselves for the most part with placing Hebrew inscriptions on small upright slabs of marble, or of common lime stone, raised after the manner generally used in the East. Many of these are broken and dilapidated; and altogether the scene offers a most desolate and melancholy appearance. And frmo the solitude of these hills, where no living creature is seen; from the ruinous state of the tombs, some broken, some overthrown, and others half open, one might imagine that the trumpet of judgment had already sounded, and that the valley of Jehoshaphat was about to render up its dead.[8]

On this consecrated ground our ancient brethren placed their lodge, as a sacred basement for the foot of the ladder, passing over a series of holy symbols, and reaching to the highest heavens; veiled from the natural eye of man by a cloudy canopy, but visible to the eye of faith as the consecrated dwelling of the Most High, seated on a throne of Light, and shining, "like jasper and a sardine stone; surrounded with a Rainbow like unto an emerald, and holding in his right hand the Great Book with seven Seals, which no man is worthy to open but the Lion of the tribe of Judah."[9]

In the lowest abyss of the valley, reminding us of the cavern in which John the Baptist dwelt in the wilderness of Judea; and the grotto of the Evangelist in the island of Patmos, where he was favoured with visions and revelations of the most sacred character, we find the basis of our symbol; an altar inscribed with the circle and parallel lines, as a support to the great lights of masonry and the theological Ladder; concerning which there have been, at various periods, several different opinions amongst the fraternity. This mysterious circle was not introduced into masonry, as a compound symbol in its present form, till about the middle of the last century. The occasion and time of its introduction are equally uncertain, but I am inclined to think that it was first inserted by Bro. Dunckerley, as a finish to the symbolism of the masonic ladder, and to form an appropriate altar for the Holy Bible, Compasses, and Square. Martin Clare's Lectures had the circle and point, but not the perpendicular parallel lines, which were a subsequent addition.

The Emblem is now sufficiently expressive to demand our utmost attention; and I have bestowed considerable pains in collecting the several interpretations which the varying fancies of ingenious brethren have at different

[8] Pict. Bibl., Joel iii., 2. Chateaub., vol. ii., p. 39. Clarke, vol. ii., c. 7. Buckingham, vol. i., p. 293. Richardson, vol. ii., p. 363.

[9] Rev. iv., 2, 3; v., 1–5.

periods attached to it; and a chronological arrangement will constitute the most intelligible and perspicuous mode of arriving at the true meaning of the symbol.

1. Its earliest reference goes as far back as the formation of the universe. The circle was supposed to represent the Deity diffused through all space, and the parallel lines, the heavens and the earth, because Moses, in recording the circumstance, commences his book with the words, "In the beginning God created the *heavens* and the *earth*." The cabalistic Jews entertained some curious fancies about this emblem, although instead of two perpendicular parallel lines, they used semicircles; but the reference was precisely the same. They held that the circle of every thing commences and terminates with God; the Almighty Creator being the *beginning* and *end* of the circle, the smallest atom within each of the semicircles proceeding from him. In the first instance it descends to the angelic, then by ordained degrees to the ethereal, from that to the lunar sphere, and then to our globe, which is first matter; this terminates *the semicircle* the farthest removed from the perfection of the Creator; then commences *the other semicircle*, which ascends to the elements; from them to mists, then to plants; from them to irrational beings, and lastly to man; ascending from a lesser to a greater intelligence, stopping the intellectual faculty of the superior intelligence of a divine origin, which is the ultimate connection between them, not alone from the angelic nature, but through it with the Supreme Divinity itself; the circle being thereby completed and graduated by all beings; that is, beginning with the heavens or celestial and superior, from major to minor; and then from minor to major, commencing with the earth or first matter, so that the circle which commences with God, and terminates with him, is completed.[10]

2. Others fancied that the circle and parallel lines referred to the earth under the influence of *night* and *day;* and that the point represented the internal fire which the Pythagoreans believed to exist in the centre of the earth. And it might also have a reference to what the Rabbins say respecting the creation, that three

[10] R. Manasseh ben Israel. Concil., vol. i., p. 3.

things were created on the first day, heavens, earth, and light; meaning by the heavens, the celestial empire; by the earth, chaos or first matter; and by light, the sovereign divine mind; the latter, under this view of the case, representing the centre.

Montfaucon[11] gives a symbol of the circle or globe divided sectionally to show its four concentric circles, attached to a figure of Isis; which he explains thus. "The first and largest circle is white, the second is blue, the third dark ash colour, the fourth red. This seems to signify the Elements. The red signifies Fire, the dark ash colour Earth, the blue Water, and the white Air. The Fire is in the centre, because it gives heat and life to all things. Here again the fire is considered the central point; and the circle is flanked on each side by two Tables placed perpendicularly, containing figures of Osiris and Anubis. Isis or the circle representing Universal nature, and the two parallel lines Eternity, and the Lord of Heaven. The above instances will show the antiquity of the symbol amongst both Jews and Gentiles.

These opinions probably originated in the account of the creation, where Moses says, "and God saw the light, that it was good; and God divided the *light* from the *darkness*." Many of the Jewish Rabbins were decidedly of opinion that by the darkness in this passage was meant the element of fire. They say that God having made it descend to that of air, it ignited and formed light. This illuminated one semicircle of the heavens, the other half being in darkness; but the light, following the rotatory motion of the primum mobile, revolving from west to east, it formed night; and then turning from east to west from whence it came, it formed day. But this light being too indistinct, for the necessary purposes of life, a more powerful agent was provided on the fourth day by the creation of the Sun.

Under this interpretation the circle represented the Earth, and the parallel lines the Sun and Moon; the former being created to rule and govern the *day*, and the latter to rule and govern the *night*, as is testified by Moses[12] and David;[13] for, as the sun is said to rule the day, because he only then appears in the firmament, so the

[11] Supplement, p. 205. [12] Gen. i., 16. [13] Ps. cxxxvi., 8, 9.

moon and stars have a delegated government in the night, because they then appear with splendour, and give a supply of light which the sun does not then immediately afford.

3. The next interpretation of the symbol refers to the creation of man. The garden of Eden contained the primary emanation of the Deity—the spirit that produced thought, reason, and understanding in the first created pair—and it was of a circular form. In the centre of this circle God placed a certain tree, which was the subject of the original covenant with his creatures; and a symbol of the life which had just been bestowed upon our great progenitors; and also of a future and still more happy life, which the circle, without beginning or end, denoted would be eternal; while the tree of knowledge of *good* and *evil* was made the test of their obedience. Adam and Eve were the two perpendicular parallel lines; being placed in the garden in a state of trial, as the objects of God's *justice* and *mercy;* and when, by transgression, the guilty pair fell from their high estate, by eating the forbidden fruit, justice demanded the threatened penalty of death, but mercy interposed, and they were banished from that happy region into a world of care, and pain, and sickness; deprived of immortality and happiness; with their eyes opened to distinguish between good and evil, and to understand the severity of their lot. And it was lest they should violate the central point by eating of the tree of life, and thus exist forever in misery, that they were expelled from the divine circle of purity, to earn their bread by daily toil and labour. This opinion was entertained by some of our brethren towards the close of the last century; and they considered the circular garden of Eden under the superintendence of our first parents, as a symbol of the Universe, which is also a circle or sphere, under the guardianship of the justice and mercy of God; and the diagram was the circle flanked by two lines perpendicular and parallel.

4. The emblem has also been referred to the Cherubic form which was placed at the gate of Paradise to prevent the return of our first parents to that region of never-ending happiness and delight, after their fall from purity and rectitude, in the attempt to acquire forbidden knowledge. The "fire unfolding itself," or globe of fire, as

described by the prophet Ezekiel, represented the Deity, and the living creatures on one side, and wheels on the other, denoted his *power* and *goodness*.[14] This was the interpretation of Archbishop Newcome, who wrote about the time when this symbol was first introduced into masonry. In his notes on the prophet Ezekiel, he gives an exemplification of our perpendicular parallel lines. In that prophet's description of the Cherubim, the following passage occurs: "They turned not when they went, they went every one straight forward;" on which the learned prelate thus remarks: "The wheels and horses of chariots bend and make a circuit in turning; but this divine machine, animated by one spirit, moved uniformly together; the same line being always preserved between the corresponding cherubs and wheels, *the sides of the rectangle limiting the whole, being always parallel*, and the same faces of each cherub always looking onward in the same direction with the face of the charioteer. This proceeding directly on, in the same undeviating, inflexible position, seems to show their steadiness in performing the divine will, which advances to its destined goal right onwards" And again, "*The axis of the former wheels is always parallel to that of the latter.* The wheels are supposed to express the revolutions of God's providence, which are regular, though they appear intricate."

5. It is a curious fact, and may serve as a practical illustration of our subject, that when the first races of men after the deluge became so numerous as to crowd the spot where they had settled with a redundant population, and they began to be afraid that it would be necessary for a great portion of them to disperse, and find out new colonies for the subsistence of their tribes, they adopted the singular expedient of building a gigantic obelisk, or tower, as a common centre to the circle which their migrations to the north, south, east, and west, might form; where they could assemble as at a point of union on any emergency that might occur; and they appear to have been desirous of embodying the principle in the peculiar figure and character of the edifice. Accordingly, it was constructed in the form of the frustrum of a cone, with a graduated ascent; and a rising

[14] Ezek. i., 4

platform, like a geometrical staircase, wound round the building, on which not only men but cattle were able to travel to the summit. The apex of the pyramid, or central point, contained an apartment secluded from common observation, which was denominated HEAVEN, and contained, amongst other secret apparatus, an observatory for astronomical purposes. And this hypothesis is corroborated by the original words of Moses, when describing the Tower of Babel, which are not, as our authorized translation specifies, "whose top shall *reach to* heaven," but "whose top shall *be* heaven;" or, in other words, the place where the autopsia of the initiations should be consummated.

Now, as this Tower contained seven stages, or apartments, one above another, it formed a stupendous illustration of the point within a circle, combined with the seven-stepped ladder, as exemplified in the Spurious Freemasonry of Persia,[15] and the Sephiroth of Jews.[16] The ascent was by an inclined plane, and therefore contained "staves or rounds innumerable," minute though they would be, with a gateway at each stage to prevent any unauthorized intrusion on the regions above. Thus Verstegan says, "The passage to mount vp, was very wyde and grete, and went wynding about on the outsyd; the middle and inward parte for the more strength beeing alle massie; and by carte, camels, dromedaries, horses, asses, and mules, the carriages were borne and drawn vp; and by the way were many lodginges and hostreries both for man and beast."[17] And Benjamin of Tudela, who inspected the remains in the twelfth century, describes it as "a spiral passage, built into the Tower, in stages of ten yards each, leading up to the summit, from which we have a prospect of twenty miles—the country being one wide plain, and quite level."

Here, then, we have an illustration, at a very early period of the world's existence, of a great Circle, with a central apartment which represented heaven, and a pathway or Ladder of graduated steps or rounds leading to it, which none but the initiated were permitted to ascend. In this apartment, at the conclusion of the initiations

[15] See the Signs and Symbols, N. Ed., p. 166.
[16] Ibid., p. 151.
[17] Rest. Dec. Int., p. 4.

according to the testimony of an ancient writer, preserved by Stobœus, and cited by Bishop Warburton,[18] "a miraculous and divine light discloses itself; and shining plains and flowery meadows open on all hands before the enraptured candidates. Here they are entertained with hymns and dances—with the sublime doctrines of sacred knowledge, and with reverend and holy visions. And now, having become perfectly initiated, they are free, and no longer under restraints; but, crowned and triumphant, they walk up and down the regions of the blessed; converse with pure and holy men, and celebrate the sacred mysteries at pleasure."

6. There are some who compare the symbol before us to the golden candlestick, flanked by two olive trees, mentioned by Zechariah;[19] the candidate representing the circle, the oil the point, and the trees the two perpendicular parallel lines. The former was an emblem of the Jewish nation, governed by the central oil, or the Holy Spirit of God; and the olive trees were the two anointed ones, viz., the King and Priest, applied by the prophet to Zerubbabel and Jeshua, who were raised up by divine providence to preside over the temporal and spiritual affairs of the Jewish nation when the Second Temple was building; and bearing an ultimate reference to the Lights and ornaments of the Christian Church.

"Upon several occasions," says Bishop Newton, "two have often been joined in commission, as Moses and Aaron in Egypt, Elijah and Elisha in the apostacy of the ten tribes, and Zerubbabel and Jeshua after the Babylonish captivity, to whom these witnesses are particularly compared. Our Saviour sent forth his disciples two and two; and it has also been observed that the principal reformers have usually appeared, as it were, in pairs; as the Waldenses and Albigenses, John Huss and Jerome of Prague, Luther and Calvin, Cranmer and Ridley, and their followers." Amongst ourselves, however, the candlestick above mentioned, or divine circle, is an acknowledged symbol of Christ, who supports the true Light, or his church, which is represented by the central point. The oil is the Holy Spirit, and the two anointed ones, or determinate witnesses to the truth and usefulness, as

[18] Div. Leg., vol. i., p. 235.

[19] Zech. xi., 4.

well as the universal application of Christianity, are generally considered to be the two St. Johns.

7. Others there are who apply the symbol to that singular type in the prophecy of Zechariah, which was intended to pourtray the establishment of the Gospel on the ruins of the Law. The passage is very remarkable, and I quote it entire. "I took unto me two Staves; the one I called Beauty, and the other I called Bands; and I fed the flock. Three shepherds also I cut off in one month; and my soul loathed them, and their soul also abhorred me. Then said I, I will not feed you; that that dieth, let it die; and that that is to be cut off, let it be cut off; and let the rest eat every one the flesh of another. And I took my staff, even Beauty, and cut it asunder, that I might break my covenant which I had made with all the people. And it was broken in that day; and so the poor of the flock that waited upon me knew that it was the word of the Lord. And I said unto them, If ye think good, give me my price; and if not, forbear. So they weighed for my price thirty pieces of silver. And the Lord said unto me, Cast it unto the potter; a goodly price that I was prized at of them. And I took the thirty pieces of silver, and cast them to the potter in the house of the Lord. Then I cut asunder mine other staff even Bands, that I might break the brotherhood between Judah and Israel."[20]

Here the circle represents the great Being who dictated the prophecy, of which himself was the subject; the centre symbolized "the flock of the slaughter," or the Jewish nation; the parallel lines, Beauty and Bands, or in other words, Love and Unity, to signify the brotherhood between Judah and Israel. Thus in a masonic song which was written about the period when this expressive symbol was introduced into the Order, we find the following characteristic reference to these two parallels under the signification of Beauty and Bands.

Ascending to her native sky,
  Let masonry increase;
A glorious pillar raised on high,
  Integrity its base.

[20] Zech. xi. 7–14.

Peace adds to olive boughs, entwin'd,
An emblematic dove,
As stamp'd upon the mason's mind
Are UNITY and LOVE.

The staff was a type of many orders of men. As a crook it was the ensign of a shepherd; as a crozier, which varies very little from a shepherd's crook, it characterises a bishop or a prophet; as a sceptre it designated a king. Bishop Hall says, in his explanation of this symbol, "the one called Beauty, was the staff of mercy, and gracious pleasurable protection; the other, called Bands, was the staff of unity for conjoining the church, or of just censure and correction to those who are opposed to it." As if he had said, since they refused to be united to me in love, under the type of the staff of Beauty, I will break my other staff, Bands, to signify the destruction of Jerusalem, and the dissolution and dispersion of the Brotherhood which hitherto subsisted between my people.

Some, however, are of opinion that the whole transaction was intended as a representation of the circumstances attending the treachery of Judas Iscariot to his Master; and, indeed, the Jews themselves expound the prophecy as being applicable to the Messiah. The application of Beauty and Bands to the science of Freemasonry was in much esteem with our brethren at the beginning of the present century; but at the reunion, being pronounced inconsistent with the general plan of the Order, it was expunged; and is now nearly forgotten, except by a few old masons, who may, perhaps, recollect the illustration as an incidental subject of remark amongst the fraternity of that period.

8. Again; the two parallel lines were sometimes referred to the two great dispensations—the Law and the Gospel, thus omitting the patriarchal system, which, as they believed, was not an actual religious dispensation, because it was not formally delivered by the Divinity unto man; and had no written Law. If this interpretation be true, the Parallels would be *Moses* and *Christ;* a disposition which does not correspond with the true principles of the Order; although in some of its definitions, the latter, as Jehovah, has been referred to the point. If the hypothesis of the Sun and Moon

be adopted, as hinted above, the parallels would be Moses and Joshua; for Onkelos, and some other Rabbins are of opinion, that Moses imparted to Joshua a part of that lustre which surrounded his countenance when he descended from the mount, after his conversation with Jehovah. Thus they say, Moses shined like the Sun, and Joshua like the Moon.

Under this head may be classed the reference which was sometimes assigned to the parallel lines, of the pillar of a cloud and of fire which accompanied and guided the course of the Israelites at their Exodus, from Egyptian bondage; and was typified in the two great pillars of Solomon's porch, that the people might recall this great event to their remembrance, whenever they went in or out of the Temple for worship; and understand that during the whole of their wanderings in the wilderness they were led, not so much by Moses as by the Hand of God, who had adopted them as his peculiar people.

9. Others have sought a solution of the enigma in the science of astronomy. These affirm that the point in the centre represents the Supreme Being; the circle indicates the annual circuit of the sun; and the parallel lines mark out the solstices, within which that circuit is limited. And they deduce from the hypothesis this corollary, that the mason, by subjecting himself to due bounds, in imitation of that glorious luminary, will not wander from the path of duty.

This doctrine will require no refutation in the opinion of those who do not consider Freemasonry to be an astronomical figment. And it is quite clear that when this symbol was introduced into the Order, the brethren never dreamt of interpreting Freemasonry by reference to the solar system, as is fully evinced by the publications of Calcott and Hutchinson, both of whom were living ornaments of the Craft at the period alluded to. But when an astronomical interpretation is determined on, whether of masonry or religion, every fact and symbol is pressed into the service, and made to coincide with the hypothesis, how strained or far fetched soever it may be. An amusing instance of this is given by Blackwell, in his edition of Mallet's Northern Antiquities, where in a note on the Scandinavian triad, he says that its members "Har, Jafuhar, and Thridi, are the three stars of Orion's belt! He also regards them as symbols of the winter's solstice, the (we presume both vernal and autumnal) equinox, and the summer solstice; an hypothesis which shows that the astronomical method of explaining ancient myths is as apt to lead learned men on a wild goose chase, as conjectural etymology."

10. In a system of Lectures used by some of the London Lodges immediately after the Union, and communicated to me at the time by an esteemed brother, a Barrister, now alas! no more; the centre and the parallels exhibited a singular specimen of pseudo symbolization which it is difficult to unravel; where the former represented the Deity, and the two latter his Justice and Mercy, as already noticed. The passage was as follows: In all our regular well formed lodges, there is a certain point within a circle, round which it is said, genuine professors of our science cannot err. This circle is bounded on the north and south by two perpendicular parallel lines. On the upper, or eastern part of the periphery, rests the Holy Bible, supporting Jacob's Ladder extending to the heavens. The point is emblematic of the omniscient and omnipresent Deity; the circle represents his eternity, and the two perpendicular parallel lines his equal justice and mercy. It necessarily follows, therefore, that in traversing a masonic lodge, we must touch upon the volume of the Sacred Law; and whilst a mason keeps himself thus circumscribed, remembers his Creator, does justice and loves mercy, he may hope finally to arrive at that immortal Centre whence all goodness emanates.

11. The elucidation of this portion of our symbol which is most prevalent in our lodge practice, at the present time, is this. In all regularly constituted lodges there is a point within a circle round which a mason cannot materially err. This circle is bounded between the north and south by two grand parallel lines, the one representing Moses, and the other King Solomon. On the upper part of this circle rests the volume of the Sacred Law of God, which supports Jocob's Ladder, and its summit reaches to the heavens; and were we as adherent to the doctrines therein contained as both those parallels

were, it would not deceive us nor should we suffer deception. In going round this circle, we must necessarily touch on both those parallel lines as well as on the Sacred Volume; and while a mason keeps himself thus circumscribed, he cannot seriously err from the path of duty.

12. One other interpretation remains to be noticed. The point is supposed to symbolize an individual mason circumscribed by the circle of virtue; while the two perpendicular parallel lines by which the circumference is bounded and supported, are the representatives of FAITH and PRACTICE. This is the definition. The point represents an individual brother, and the circle is the boundary line of his duty to God and man; beyond which he ought never to suffer his passions, prejudices, or interests to betray him. The two parallel lines represent St. John the Baptist and St. John the Evangelist, who were perfect parallels in Christianity as well as in masonry; and upon the vertex rests the Holy Bible, which points out the whole duty of man. In a progress round this circle, the two lines and the Bible restrict us to a certain path, and if this path be steadily persevered in, it will enable us to mount the ladder, through the gates of Faith, Hope, and Charity; and finally to take our seat in the blessed regions of immortality.

In the lectures which are still delivered in our old provincial lodges, the following illustration is used. From the building of the first Temple at Jerusalem to the Babylonish captivity, the lodges of Freemasons were dedicated to King Solomon; having from the deliverance out of Egypt to the first named period been dedicated to Moses. From the building of the second Temple to the advent of Christ, they were dedicated to Zerubbabel; and from that time to the final destruction of the Temple by Titus, they were dedicated to St. John the Baptist. Owing to the calamities which were occasioned by that memorable occurrence, Freemasonry declined; many lodges were broken up and the brethren were afraid to meet without an acknowledged head. At length a secret meeting of the Craft was holden in the city of Benjamin; who deputed seven brethren to solicit St. John the Evangelist, who was at that time bishop of Ephesus, to accept the office of Grand Master. He replied to the deputation, that having been initiated into masonry in his youth, he would cheerfully acquiesce in their request, although now well stricken in years; thus completing by his learning what St. John the Baptist had begun by his zeal; and drawing what Freemasons call a line parallel; ever since which, the Lodges in all Christian countries are, or ought to be, dedicated to the two St. Johns.

These various conjectures, like "a cluster of pomegranates with pleasant fruits," which have been the produce of different periods and phases of the Order, are all ingenious if they be not orthodox. Like the fat kine of Pharoah, they equally display a beneficial nurture, and point out its moral and religious tendency. Here is no overstraining of facts, no unnatural antagonism, to serve the purpose of some wild or untenable theory; but every interpretation is alike consonant with the deductions of reason, without being at variance with revelation, or contrary to the established laws of Nature. Each, like the growth of the acacia, has budded in its spring, flourished its brief period of summer, and shed its leaves in autumn, to make room for its successor; which has pursued a parallel course; and the following lecture will be devoted to a consideration of which is the most eligible interpretation in consonance with the general principles on which Freemasonry has been founded.

LECTURE VI.

Epistle Dedicatory

TO

BRO. THOMAS CLEAR,	W. M.
— J. C. SMITH,	S. W.
— G. T. CASWELL,	J. W.
— C. S. CLARKE,	P. M.
— JOHN CRITCHLEY,	TREA.
— G. PRICE,	SEC.
— T. BOLTON,	S. D.
— D. L. DAVIS,	J. D.

Of the St. Peter's Lodge, Wolverhampton, No. 607.

MY DEAR BRETHREN AND FRIENDS

The consciousness that you still retain a lively recollection of my presence amongst you, when a mutual interchange of affection and thought cemented an intercourse which commenced under circumstances of the greatest personal interest to myself; and was not only continued during the entire period of my residence as the Incumbent of the Collegiate Church, and the head of the Ecclesiastical Establishment in the town, but terminated in a public and spontaneous demonstration of the feelings that you were kind enough to entertain towards me during a series of trying events, cannot fail to excite in my bosom sensations of the most pleasurable nature, accompanied by a lively gratitude for the continuance of that friendship which sustained no diminution amidst the hostile denunciations of a clique of interested individuals who were leagued in an unnatural coalition to ruin my peace of mind at the least, if they should fail to accomplish a more destructive purpose.

During the arduous conflict, your sympathy consoled me,—your friendship animated me,—and ultimately, your assistance procured for me a complete and unquestioned triumph. I have much pleasure in having this public opportunity of assuring you, that I entertain no unkindly feelings against those whose hostility was most bitter. Freemasonry has taught me a different lesson; and I sincerely tender my unsolicited forgiveness to them, in Christian charity, with the same cordiality as I hope myself to be forgiven at the bar of judgment. The sole aim of my life has been to benefit my fellow creatures; and my principles are well embodied in the following lecture which I have the honour to dedicate to you.

If the two parallel lines by which the circle and point are flanked and supported, have, as I believe, a reference to faith and practice, they include forgiveness of injuries; and I trust that my practice will always verify this valuable principle of masonic teaching; that when I meet my persecutors at the last great tribunal, I may salute them as friends with the grip of a Master Mason; and, by the five points of fellowship, unite with them in an indissoluble chain of sincere affection, which may continue unbroken throughout all eternity.

Believe me to be,

Worshipful Sir,

And esteemed Brethren,

Your faithful friend and Brother,

GEO. OLIVER, D.D.,

Honorary Member of St. Peter's Lodge.

SCOPWICK VICARAGE,
November 1, 1849.

Lecture the Sixth

Enquiry into the true meaning of the Circle and Parallel Lines.

"In regard to the doctrine of our Saviour, and the Christian revelation, it proceeded from the East. The Star which proclaimed the birth of the Son of God, appeared in the East. The East was an expression used by the prophets to denote the Redeemer. From thence it may well be conceived that we should profess our prayers to be from thence; if we profess, by being masons that we are a society of the servants of that Divinity whose abode is in the centre of the heavens."

HUTCHINSON.

"To Thee, whose temple is all space,
Whose altar, earth, sea, skies!
One chorus let all being raise!
All Nature's incense rise!

POPE

So MANY reflections arise in the mind on a full consideration of this comprehensive symbol; like a majestic river augmented by the rich contributions of its tributary streams; and so various are the opinions which successive races of masons have entertained respecting its real interpretation, that we find it impossible to dismiss the subject without some brief statement of our own views on those particular points which have hitherto divided the fraternity. We need be under no surprise that interpretations of an abstruse symbol should vary by passing through different hands, because every one knows that in the most common transactions of life, accounts related by several eye-witnesses, however they may assimilate in facts, differ considerably in the details.

This is not a novel observation, for it is mentioned by Sir Walter Raleigh and many others, as constituting one of the greatest obstructions in the compilation of history. Different authorities produce such adverse accounts of the selfsame fact, that confidence is shaken, and it becomes difficult, if not absolutely impossible, to distinguish between truth and falsehood. Such being the obstacles which are always found to embarrass a narrator of historical truth, we no longer wonder that there should be diverse opinions on subjects merely speculative, which are not based on any indisputable authority.

Such is the position of the symbol before us. It has been interpreted differently at different periods, and successive Grand Lodges have sanctioned each new hypothesis as it arose; although in some cases the explanation appears to have been at variance with analogy and the general principles of the Order. The theory, for instance, which places the Deity in the centre of the circle, is opposed by his own assertion when He says, "Do I not fill heaven and earth, saith the Lord."[1] The point is evidently a misappropriation of the Creator; and he was more correctly represented in the ancient hieroglyphics by *the entire circle*. The cabalistic Jews, indeed, used a symbol resembling the circle and point, by describing a circle round the letter Jod; but it was the *letter*, and not the situation where it was placed, viz., in the centre, which they interpreted as an emblem of the Deity; which might probably be the reason why Hutchinson, in our motto, confines him to the centre of heaven, when our Scriptures represent him as being peculiarly present, not in the centre, but in the *highest* heaven.

This is further evidenced by the fact, that they used an equilateral triangle for the same purpose more frequently than a circle. It was the Letter that formed "the idea of God;" and they pronounced upon it this glowing enconium. "It is a ray of Light which darts a lustre too transcendent for the contemplation of mortal eye; and though the thoughts of man may pervade the universe, they cannot reach the effulgent light which streams from the letter Jod."

The centre is a symbol of Time, and the circle of Eternity. The latter, like the universe, being unlimited

[1] Jer. xxiii., 24.

in its extent; for time is but as a point compared with eternity, and equidistant from all parts of its infinitely extended circumference; because the latter occupied the same indefinite space before the creation of our System, as it will do when time is extinguished, and this earth, with all that it contains, shall be destroyed. And therefore the hypothesis which would confine the Eternal, whom the heaven of heavens cannot contain,[2] to the emblem of time, is evidently unsound, and must be rejected. The ancients took the entire Universe for the centre, and left the circumference undefined; but still it was tenaciously asserted that *deus circulus est*, including the circumference how boundless soever it might be; and our emblem was not intended to embrace such an indefinite and comprehensive illustration.

It is true, in the Pythagorean circle, or the universe, the central fire represented Unity; but it was not referred to the One God the Creator, but to Vesta, of whose inextinguishable fire it was an emblem, for Vesta signifies fire; and therefore her temples were generally spherical, to represent the universal circle. The Basi deans, on the other hand, confined their supreme Deity, whom they called Iao (Jehovah), Abraxas or Meithras, to the circle of the year, which is as bad as placing him in the centre, and perhaps worse, as a single year is a more perverse limitation than all time. St. Jerome informs us that Basilides gave to the Almighty the monstrous name of Abraxas, because, according to the import of the Greek letters, and the number of days in the Sun's course, Abraxas is found in the circle of the Sun, in the same manner as the word Meithras was also found by the Gentiles, because the letters contained the same number. Thus Basilides made the circle to consist of three hundred and sixty-five heavens, and placed Abraxas or the Sun in the centre; and referred him to Jesus Christ the Sun of Righteousness.

In conformity with this doctrine, we find on numbers of the Basilidean gems, the figure of a Cock; which. according to Montfaucon, "is, without doubt, a symbol of the Sun, who holds in his hand a whip to animate his horses with, while travelling on his unvarying circle. and

[2] 1 Kings viii., 27.

has probably a cock's head given him, because that bird generally proclaims the Sun's rising. And it is to be observed that all the gems of this character containing figures of Abraxas, have generally relation either to the Sun or its operations, as most other Egyptian figures have. Many of these ancient heretics believed that Jesus Christ was the same with the material Sun; which notion gave occasion to their mixing Christianity with the divinities of that superstitious nation."[3] Montfaucon has given seven folio engravings containing several hundreds of these gems; many of which contain the letters *A* and *Ω*, to signify the eternity of God.[4]

The Deity, however, is Omnipresent, and cannot be confined to any individual locality, but is equally diffused throughout the entire universe. Grand Master David asks, "whither shall I go from thy Spirit; or whither shall I flee from thy presence?" And answers, "if I climb up into heaven thou art there; if I go down to hell, thou art there also. If I take the wings of the morning, and remain in the uttermost parts of the sea, even there also shall thy hand lead me, and thy right hand shall hold me." It is true, that at one period the light of God's knowledge shone only on the favoured land of Judea, while all the rest of the world were involved in the darkness of idolatry;—but God was even in the darkness; for the pious Psalmist continues, "the darkness is no darkness with thee, but the night is as clear as the day."[5]

The Jewish doctors contended that their country was the only region of true Light; and that Jerusalem constituted the central point of the Earth, which they regarded as an extended plain; for it had been asserted by their inspired monarch, that God wrought out his salvation, by establishing them and their religion "in the midst of the earth."[6] Macrobius tells us that *veteres omnem habitabilem terram extensæ chlamydi similem esse dixerunt.* And we have the testimony of several early Christian travellers to the existence of this persuasion. Thus bishop Arculf, whose journey was accomplished in the year of Grace 700, says that "near to Golgotha he

[3] Montf. Ant., vol ii., p. 227. [4] See the F. Q. R., 1848, p. 378
[5] Ps. cxxxix., 6–11. [6] Ps. lxxiv., 12.

observed a lofty column which at mid-day, at the summer solstice, casts no shadow, which shows that it is the centre of the earth." Bernard, who travelled in the year 867, speaks of the same thing. "Between the aforesaid four churches is a parvis without roof, the walls of which shine like gold, and the pavement is laid with precious stones; and in the middle four chains coming from each of the four churches, join in a point which is said to be the middle of the world." Saewulf, who travelled A.D 1102, adds, "at the head of the church of the Holy Sepulchre, in the wall outside, not far from Calvary, is a place called Compass, which our Lord Jesus Christ himself signified and measured with his own hand as the middle of the world." Sir John Maundeville, A.D, 1322, adds his testimony to the same belief. He says, "Judea is the heart and middle of all the world; and hence it was right that he who created all the world should suffer for us at Jerusalem, which is the middle of the world, to the end and intent that his passion and death, which was published there, might be known equally to all parts of the world." And speaking of the church of the Holy Sepulchre, he says, "in the midst of that church is a compass, in which Joseph of Arimathea laid the body of our Lord when he had taken him down from the cross, and washed his wounds. And that compass, men say, is the middle of the world."

The Rabbi Judah à Levi says, in confirmation of this hypothesis, "as the heart is in the centre of the body, so is the Holy Land the centre of the world's population. and is therefore more acceptable to the Lord. For as the world is divided into seven climates, that land is situated in the best of them; the Psalmist describes it as beautiful for situation, and the joy of the whole earth. There it was that Cain and Abel contended, and Cain's punishment consisted in being banished from it. The patriarchs selected it for their place of interment. Abraham satisfied the children he had by Keturah with presents, that they might quit it and leave Isaac in quiet possession of its fertile mountains and plains. Esau abandoned it entirely to Jacob, and went to dwell in Seir. From all which it is collected, that this country was ever considered to be peculiarly and exclusively holy." Hence the Holy Land was esteemed the central point, while the rest of the world occupied the outer circle, and were accounted by the Jews to be profane.

This interpretation, however, is insufficient to reconcile the anomaly of confining the Deity to the centre, or any other place; although his divine Shekinah might and did occupy the Sanctum Sanctorum, in the Temple at Jerusalem, which, as we have just seen, was believed to constitute the centre of the earth.

The true religion, or Light, as we Christians conscientiously believe, is now universally diffused, and therefore would be more aptly represented by the circumference, bounded only the limits of the habitable globe; for the light is gradually dispelling the darkness by the labours of our indefatigable missionary establishments; and in God's good time it will enlighten the whole universe, radiating in all directions to the heaven of heavens, till it blends with the circumambient lustre which surrounds the throne of God.

It would be the height of presumption to circumscribe the omnipresent Deity within the narrow limits of a religion, which, though revealed from Heaven, and consequently true, during the period of its lawful authority, is no longer in force. The Sceptre has departed from Judah, and the temporary system which it upheld has been superseded by a dispensation that is destined to be universal. The Jewish religion was for one nation only, while Christianity is a light to lighten the Gentiles, as well as the glory of the people of Israel. Still we firmly believe that the Old Testament is not contrary to the New; for both in the Old and New Testament everlasting life is offered to mankind by Christ, who is the only Mediator between God and man.

Let it not be thought that it is my intention to offer any disparagement to our Jewish brethren, or to the religion they profess; for the subject I am discussing is open to them, and to all who acknowledge the being of a God. I interpret the symbol according to my own convictions as a minister of the Gospel who is not ashamed of the Cross of Christ. The Jews are an extraordinary people, for they remain unchanged amidst the political fluctuations of nearly 2000 years. They are entitled to our consideration and our gratitude; for we are indebted to them for the blessing of Moses and the prophets, without which we should be ignorant of the history of the world, and the ways of God to man; and should have wanted those glorious and divine precepts which are a source of consolation when almost everything else fails. Homer and Virgil are sublime; but what are they when compared with Moses and David, who were inspired by the Deity to indite a series of divine hymns which cast every other composition into the shade. They are full of such exalted strains of piety and devotion, as a writer of the last century well expresses it, such beautiful and animated descriptions of the power, the wisdom, the mercy, and the goodness of God, that it is impossible for any one to read them without feeling his heart inflamed with the most ardent affection towards the Great Creator and Governor of the Universe.

And although, in conformity with the voice of their own prophet, they are scattered among all people from one end of the earth to the other, yet they are not totally destroyed, but still subsist as a distinct people. "The Jewish nation, like the Bush of Moses, hath been always burning but never consumed. And what a marvellous thing it is that after so many wars, battles, and seiges; after so many fires, famines, and pestilences; after so many years of captivity, slavery, and misery, they are not destroyed utterly, and although scattered are still distinct. Where is there anything comparable to this to be found in all the histories, and in all the nations under the Sun?"

As to the project of converting the Jews to Christianity—it is not to be thought of until the time of their restoration shall arrive. The attempt is sure to be unsuccessful, because it bears a close resemblance to a wish to frustrate the prophecies. It has been predicted that they shall be dispersed among all nations and not be amalgamated with any; but their conversion would produce marriages and intermarriages with Christian people, and in a few generations the distinction would be lost. God has appointed a time when they shall be restored and converted to the true religion; and when that time comes, nothing can prevent the fulfilment of the prophecy. The omnipresent Deity wants no assistance in the accomplishment of his purposes; and when it is his gracious pleasure, there shall be but one fold under one shepherd, and the Messiah of Jews and Christians shall reign for ever and ever.

Dr. Schulhoff, in a speech recently delivered at Birmingham, observed, "we cannot pray without adopting the language of the Jews; we cannot read the Scriptures without meeting with their history upon every page. Men were sent into the world as a question, a riddle, or an enigma, not yet answered or resolved, and for this reason they might ask who were the Jews? They were that little family who were born in Canaan, who went into Egypt, who had been hunted from country to country, from land to land, from dungeon to dungeon, from the death by fire to the death by water. Who were the Jews? Ask the dust of Canaan; ask the walls of Zion, overcome by the scimitar of the Saracen. The Jewish was the nation out of which, according to the flesh, came the Messiah; which produced the Apostles, and the early Christian converts." And he concluded by asserting that *the present position of the Jews is necessary for the honour of God; and is a great proof of the truth of the Gospel.*

Many of the heathen nations who entertained imperfect notions of God's universal presence, as in the masonic definition above noticed, confined him to the centre. The first settlers in Egypt transmitted to their posterity an exact copy of our point within a circle, expressed in emblematical language. The widely extended Universe was represented as a circle of boundless light, in the centre of which the Deity was said to dwell; or in other words, the circle was symbolical of his *eternity;* and the perpendicular parallel lines by which it is bounded, were the two great luminaries of heaven, the Sun and Moon, the former denoting his *virtue*, the latter his *wisdom.* And this idea was generally expressed by the head of a hawk placed in the centre of a circle, or an endless serpent enclosing an eye. In like manner Pythagoras considered the central fire to be the mansion of the Deity or God; and assimilated it with the monad, because it is the beginning and ending of number. In the Stone temples of the Celtæ we find indications of a similar belief. But divine revelation has taught us a more correct and rational doctrine; and we possess the advantage of knowing that Jehovah or God in the Old Testament, is no other than the Messiah or Christ in the New. Hence the following singular elucidation of the above doctrine has been deduced, and is actually in practice by some of our lodges at the present day.

Our ancient brethren, in depicting on the Tracing Board the Plumb, the Rule, the Level, and the setting Maul, intended by the latter to represent the point within a circle; and in a speculative sense referred them to the death of T G A O T U, or Christ upon the Cross. And as, in the allegory, the above instruments were used, so in the fact, they were really concerned in the death of the actual Grand Master, whose Cross was formed from the junction of the Level and the Plumb. Now as the point within a circle is a very ancient symbol, and was venerated in many nations, as a representative of the Deity, it is considered by those who adopt this exposition to refer to the Eternal Father, who gave his Son to die for the sins of men.

They go on to say in their application of the symbol, that as Speculative Masons we ought not to be contented with deriving one moral lesson from every single emblem depicted on our Tracing Board; but to consider each as a text on which to build a copious Lecture. For instance; the point within a circle, in the first degree, has an application totally different from what it bears in the third. In the former it describes the boundary line of a mason's path through this life, the limits of which are the precepts of the Law and Gospel, by which he is circumscribed to prevent his going astray. And if he adheres to these precepts, amidst evil report and good report, the Ladder, whose foot is placed on the Holy Bible, will conduct him to a celestial mansion which is at present veiled from mortal eye by the starry firmament.

But in the third degree this emblem has a mystical reference to Christ as our centre; according to his own declaration—"where two or three are gathered together in my name, there am I *in the midst* of them." The prayer which was formerly used is considered very appropriate by those who adopt this mode of explaining the circle and point. It began thus; "O Lord God, thou great and universal mason of the world, and first builder of man as it were a Temple, be with us as thou hast promised, that when two or three are gathered together in thy Name, thou wilt be in the midst of them, &c."

After our Lord's resurrection, the disciples changed their time of assembling together in closely tyled lodges, for fear of their enemies, from the seventh to the first day of the week; and on that day our Lord appeared to Mary and directed her to go to his brethren, and inform them that he was about to ascend to the Grand Lodge above, into the presence of him who was both his Father and their Father; and in the evening when they were assembled together, "the doors being shut," or in other words, the lodge being closely tyled, came Jesus and stood *in the midst* of them, making use of the masonic greeting, "Peace be with you." Our brethren would naturally feel surprised at the presence of a stranger in a closely tyled lodge; but when he had given them proofs (by showing them those signs of distress in his hands and feet and left breast,) that he was their Brother, they dismissed all fear and rejoiced exceedingly.

It was here that he promised to be always in the midst of them; and cheered by the recollection of this gracious promise, they were naturally led to the hope of finding him within the centre of their circle whenever regularly assembled in a just and perfect lodge dedicated to the glory of God and the welfare of mankind. Hence all Christian masons are required by the circle and point to understand, that whenever they meet as brethren, his Allseeing Eye is present in the centre; and thus they are incited to discharge their duties towards him and to each other with freedom, fervency, and zeal. Thus, as those whose who use this illustration conclude, all our emblems having a tendency to inculcate the practice of virtue amongst its genuine professors, the more moral lessons we can derive from any of our emblems, the more securely founded will be the respect which Masonry may claim at the hands of all mankind.

The two perpendicular parallel lines have been appropriated to Moses and Solomon, on the presumption that they represent those two parallel edifices, the Tabernacle and Temple; these being the only two places in the early ages of the world where the true worship of God was celebrated; and the above masonic worthies being the builders of these mysterious sacred oratories, are considered to be the legitimate patrons of an Order that professes to be based on the science of decorative architecture.

Some of our brethren, however, are inclined to question the propriety of this ascription; and to doubt whether its truth can be accurately demonstrated by a reference to facts. They argue that if one of these parallels be assigned to Moses as the builder of the Tabernacle, the paternity of the other may be justly contested by Zerubbabel and Herod, each of whom built the Temple at Jerusalem, as well as Solomon. And if *that* edifice be meant, which, according to the testimony of the prophet Haggai, was the most glorious, the preference must be given to Herod, because it was his Temple which was enlightened by the actual presence of the Prince of Peace, of whom the Shekinah of the Jews was but a symbol; and consequently was superior to that of either Solomon or Zerubbabel.

The true mason, however, will hesitate to admit the claim of Herod, as one of the Parallels of the Order; because he was so far from professing the true faith, that the historian has recorded of him, that he laboured zealously to remove all the prejudices of the Jews in favour of the law of Moses, by introducing among them the customs of heathen nations; by which he intended, if possible, to Romanize Judea. The designs which he had manifestly formed against their religion, and his violation of every custom to which the Jews were sincerely attached, appear to have been considered the certain forerunners of some dreadful evil to that people. Herod was in name their king, but in deed the enemy of their country and their God.

The above appropriation is further encumbered with the anomaly of four claimants to the parallelism of the Order; viz., Moses, Solomon, Zerubbabel, and Herod; and I cannot see any alternative but to admit or reject them all; and those who have adopted this view of the case, unanimously agree to prefer the latter. If, however, as some are inclined to think, the parallels be esteemed symbols of the two great dispensations of religion, they would then apply to Moses and Christ, who were really parallels in many important respects, while Moses and Solomon possessed no single attribute in common, which can give a colour to the position in which they have been placed as equal patrons of Masonry, and joint supporters of the circle and point.

The two former coincide in character and attributes in many remarkable particulars; and there was no other prophet who ever resembled Moses, so much as Jesus the Messiah. None of the ancient prophets can answer this description. None of them were law-givers like Moses; none of them had such clear communications with God; for their prophecies were revealed to them in visions and dreams. Moses and Christ are the only two who perfectly resemble each other in these respects. The Jews were addicted to the idolatry of the Egyptians; and were taught by Moses the true way of worshipping God. Subsequently they were the slaves of superstition; when Jesus Christ taught them true religion. The system of Moses was confirmed by stupendous miracles, so was that of Christ. Moses led the people to the promised land; and Christ directs us to a better country. Moses fasted forty days, and so likewise did Christ. Moses and Christ equally fed the people miraculously. Moses led the people through the Red Sea; and Christ walked upon the sea. The face of Moses was surrounded by a bright glory when he descended from the mount; and the face of Christ shone like the sun, upon Mount Tabor. Moses deputed twelve men to survey the land; and Christ committed the same office to twelve Apostles.

St. John mentions the Christian parallelism between Moses and Christ in express terms, when he says, "the Law was given by Moses, but Grace and Truth came by Jesus Christ;"[7] intimating that although the resemblance between Moses and Christ was striking, yet there was no absolute equality; for that the latter was as much superior to the former, as Grace and Truth are superior to the Law; and St. Paul tells the Jews that they are not now under the Law, but under Grace;[8] adding in another place, that by Christ all that believe are justified, which they could not be by the Law of Moses.[9] And he more plainly asserts in another place, that "Moses was faithful as a *servant*, but Christ as a Son;" and that Christ was counted worthy of more

[7] John i., 17. [8] Rom. vi., 14. [9] Acts xiii., 39.

glory than Moses, inasmuch as "he who hath builded the house hath more honour than the house;" that is, the difference between Christ and Moses is that which is between him who creates and the thing created; and then, having before ascribed the creation of the world to Christ, he adds, "he that built all things is God."[10]

Still Masonry has not ventured to introduce the Redeemer of mankind as one of its great Parallels, because in neither of his natures has he any equal. As man he was sinless—as God he was divine. Besides, Christ is no other than Jehovah, T G A O T U, and he is symbolized by the circle. If, therefore, the parallel lines have any reference to this great Being, they can be no other than his divine and human natures, which would be masonically inapplicable; and we have already a very significant emblem to the same effect, viz., the pentalpha, double triangle, or seal of Solomon.

The circumambulation of the candidate is generally presumed to have an intimate connection with the symbol under our consideration; and therefore some have considered that originally the space included within the two parallel lines, from one extremity to the other, north and south as well as east and west, which is an oblong square, represented the Lodge; the circumference of the circle being the boundary line threaded during the ceremony; the centre being the candidate, or rough ashlar, the point from which all superficial and solid figures emanate, in the first or most superficial part of Masonry; for if one limb of the compasses be fixed, and the other movable, the point formed by the first touch of the latter, if continued, would form the circumference of a circle. In this case, the pedestal on which the Sacred Volume rests would represent Wisdom, or the W. M.; and the two lines Strength and Beauty, or the S. W. and J. W.; the Ladder, the three Theological Virtues, and the summit, perfection, symbolized by the perfect ashlar. The whole being crowned by an ethereal mansion veiled from mortal eye by the starry firmament; or, as it is termed by Job, "the face of God's throne,"[11] and surrounded by clouds and darkness,[12] that though the dwelling-place of the Most High is hidden from our

[10] Heb. iii., 3-6. [11] Job xxvi., 9. [12] Ps. xcvii., 2.

view, his decisions are the result of unerring justice and perfect truth.

This constitutes a lucid symbol of the omnipresent and omnipotent Deity, whose Throne is in the highest heavens, the region of perpetual light, and not in the central point of Time: for the act of going to heaven is always represented by ascending.[13] Job says "God is in the *height;* thick clouds are a covering to him;"[14] which is confirmed by Jeremiah, who adds, "the Lord shall cry from on *high*,"[15] and not from the centre.[16] David in his remarkable prophecy says, "Thou hast *ascended* up on high."[17] And in fulfilment of this prediction, the Apostle affirms that Christ was made higher than the heavens;[18] ascended above all the heavens;[19] and sitteth at the right hand of Majesty in high places."[20]

This idea of perfect happiness above the skies, the residence of the Supreme Deity, was not confined to the Jews and Christians, but was universally admitted by the heathen. It is clearly enunciated by Homer in the Iliad, and by Hesiod in his Theogony, who measures the distance between the highest heaven and the lowest hell, by the falling of a heavy weight, which he says, notwithstanding the inconceivable rapidity of its course, will take nine days and nights to fall from heaven to earth, and the same length of time to descend from the earth to Tartarus. It will, however, be observed, that although God is thus said to dwell in the highest heavens, yet being omnipresent, he is bound to no limit or space; and the expression is merely intended to imply that there his glory is more particularly manifested.

Considered in the above point of view, the figure under our notice constitutes one of the most glorious and expressive symbols that has ever been submitted to the consideration of mankind.

The most reasonable opinion which appears to have been formed on the circle and point, is that which makes the candidate represent the centre, placed within the

[13] See Rev. xxi. [14] Job xxii., 12–14. [15] Jer. xxv., 30.

[16] Consult Ps. cxlviii., 1. Heb. vii., 26. Eph. iv., 8, 10, and many other passages.

[17] Ps. lxviii., 18; and compare John iii., 13, with Eph. iv., 9, 10.

[18] Heb. viii., 26. [19] Eph. iv., 10. [20] Heb. i., 3.

circle of duty, and his conduct regulated by the twc lines of *faith* and *practice;* because, without the operation of these heaven-born qualities, it will be impossible for him to keep his passions within the boundary line of reason and Scripture, and to avoid those errors which will interrupt and retard his progress through this world to another and a better. The same idea was symbolized by our brethren of the last century, by the figure of a brother properly clothed, standing beneath the extended legs of a pair of compasses.

If, then, the two parallel lines represent the faith and practice of a rational soul, they are with the strictest propriety assigned to the two St. Johns, because the one finished by his learning what the other began by his zeal. The Evangelist was the most perfect personification of *faith* that the world ever witnessed; not only from the tenor of his writings, and because he was the beloved disciple of his Master, but also because his faith preserved him as witness, and the only witness amongst the Apostles of Christ, of the destruction of the Jewish polity, and the triumph of that universal dispensation which shall ultimately, as I sincerely believe, supersede all others, and cover the whole earth as the waters cover the sea. And in like manner, the Baptist was the personification of *practice*, because he confined his preaching to repentance and good works, both of which are exclusively practical. And hence it is believed that these two great and pious men acquired the distinguishing character of the patrons of Masonry.

I have given the reasons which have been assigned for parallelism in another place,[21] but, as every book ought to be perfect in itself, it may be necessary to repeat them here. "The two St. Johns were equally distinguished by the Redeemer of mankind; the one as a burning and shining light—whence the gnostics assumed that the Spirit of light entered into John the Baptist, and therefore that he was in some respects to be preferred to Christ—and the other was called the beloved disciple, and the divine." Thus they formed a personification of Greatness and Goodness, which were ever the qualities that drew down public respect and applause; and amongst

[21] See Mirror for the Johannite Masons, p. 114.

heathen nations, as we learn from Selden, constituted the attributes of the celestial deities, and elevated deceased mortals to the skies.

For these reasons, the two St. Johns were likened to the pillar of *fire* and *cloud* which attended the Israelites in their escape from Egyptian bondage. The Baptist, because he preached the unquenchable *fire* which is the punishment of sin; and the Evangelist, because he inculcated the subdued virtue of brotherly love, the practice of which, like the operation of the *cloud* to the camp of Israel, when it moderated the heat of the sun in that parched climate, would serve to avert the ever-burning fire of hell.

This pillar was a *light* and a *guide* to the Israelites through the wilderness of Sin, that they might attain the Promised Land in safety; and the two St. Johns—one by announcing the Saviour, and the other by his benevolent doctrines—are a light and a guide to all mankind while toiling through the sinful wilderness of this world, that they may arrive at the heavenly Canaan, and rest for ever from their labours. Besides, St. John the Evangelist was chosen to be a witness of Christ's transfiguration, and was actually enveloped in the cloud on that remarkable occasion.

Again, as the columns of Solomon's porch, called Jachin and Boaz, were typical of this cloudy and fiery pillar, so the early Christians likened them to the two St. Johns, which appears, in the estimation of our ancient brethren, to have made parallelism complete. Boaz represented strength, and Jachin to establish; and together, they referred to the Divine Promise, that God would establish his holy Temple in strength. The former referred to the Sun, which rejoiceth as a giant to run its course;[22] and the latter the Moon, because, like the pillar of a cloud, its light is mild and beautiful, being only a reflection of the Sun's more powerful rays; and hence it was prophesied of Solomon, that his kingdom should remain in peace and righteousness so long as the Moon endureth.[23] The promise of David includes both. "His seed shall endure for ever; and his seat is like as the Sun before me. He shall stand fast for evermore as

[22] Ps. xix., 5.

[23] Ibid. lxxii., 7.

the Moon, and as the faithful witness in heaven."[24] Hence, according to the testimony of Jarchi, Solomon said: "My kingdom being thus permanently established as the *sun* and *moon*, its duration shall be marked by the existence of these pillars, for they will remain firm and immovable as long as my successors shall continue to do the will of God."

In like manner the two St. Johns were esteemed pillars of Christianity, and patrons of Masonry. The one representing *strength*, and the other a principal agent to *establish* the permanency of both by inculcating brotherly love or charity, which is their chief virtue, and of more value than all the rest. By these instruments Christianity and Masonry have been established in such strength, that they will endure for ever. And at that period, when the designs of Omnipotence are completed, the Sun and Moon, by unmistakable tokens, shall declare to the world that their glory is expiring. The Sun will turn into darkness, and its light being thus withdrawn, the Moon will be obscured; at which period St. John the Baptist, as a righteous man, will shine forth as the Sun, standing at the left hand of the Judge amidst the clouds of heaven; while the pure and holy doctrines of his illustrious parallel will for ever remain as the employment of saints and angels in the heavenly mansions of the blessed; where there is "no need of the Sun, neither of the Moon to shine in it, for the glory of God will lighten it, and the Lamb is the light thereof."[25]

According to the opinion of Dean Stanhope, who is no slight authority in such matters, the office of John the Baptist consisted in promulgating the very doctrines which distinguish the noble Order of Freemasonry. He employed himself in "making guilty people sensible of their sins, reproving open wickedness, unmasking hypocrisy, beating down spiritual pride; importuning men to repentance, by representing, with a faithful zeal, the horrible mischiefs and dreadful conclusion of a wicked course of life, and the terrors of that Master, who, at his coming to purge the floor, will not fail to separate most nicely between the wheat and the chaff, and burn the latter with unquenchable fire. We shall do well to take

[24] Ps. lxxxix., 36, 37. [25] Rev. xxi., 23.

the Baptist for an example of our conduct, by living a life of severe virtue; by boldly rebuking vice; and if by this we incur the displeasure of men, by suffering with a constancy like his. If this were duly attended to, a mighty change would soon be effected even in the profligate and profane world."

Such investigations as these constitute the true poetry of the Order, and render the pursuits of Freemasonry of real and intellectual value to the intellectual man. Whoever, therefore, is desirous of regulating his life and conduct by the teaching of Freemasonry, will do well to make its symbols his study, and endeavour to bring their hidden meaning to bear upon the every-day occurrences of life. Plutarch has somewhere told us that while Alexander the Great was but a boy, so far from rejoicing at his father's success in battle, complained to his young companions that his father would leave nothing for him to do when he grew up to man's estate. They replied, that all which had been done by his father was for his enjoyment; but he said, what better shall I be in possessing ever so much, if I do nothing? So it is with us. Freemasonry has done a great deal, but it has left something for every individual Mason to do; and he who does it not, must not expect to be rewarded.

Nothing can be a greater anomaly than knowledge without practice. It is like hiding our talent in a napkin, or placing our light under a bushel. To produce a beautiful effect, the talent must be used and the light displayed, or we shall be pronounced unprofitable servants, and cast into outer darkness; which will be a most lamentable destiny for those who boast that they have been brought to light, and have consequently enjoyed superior advantages in acquiring information on which to found a concurrent practice. But where the central point has been illuminated by the bright rays proceeding from Eastern wisdom, and invigorated his faith by the practice of moral virtue, he will gradually ascend the innumerable rounds of the Masonic Ladder, and enter into peace when the archangel shall proclaim that time shall be no more.

LECTURE VII.

Epistle Dedicatory

TO

BRO.	C. E. ANDERSON,	W. M.
—	G. POWELL,	S. W.
—	J. G. SHIPWAY,	J. W.
—	T. DOUGLAS HARRINGTON,	D. P. G. M. & P. M.
—	REV. W. T. LEACH,	CHAP.
—	J. WHITLAW,	TREA.
—	W. P. STREET,	SEC.
—	N. RYAN,	S. D.
—	W. N. COURTNEY,	J. D.
—	R. CHALMERS,	D. OF CER.
—	J. V. NOEL,	STEWARDS,
—	J. M'COY,	

Of St. George's Lodge, Montreal, Canada, No. 643.

DEAR BRETHREN,

The circumstances under which I was elected an honorary member of your Lodge, have afforded me unfeigned pleasure. If I have rendered any services to Masonry, as you are pleased to say, by my publications, I assure you that the time which has been employed in their composition passed very agreeably to myself, because one of the principal amusements of my leisure hours has ever been the study of antiquity, and the acquirement of hieroglyphical knowledge. I am therefore doubly rewarded, inasmuch as you have added to the pleasure which such researches have conveyed to my mind, an unequivocal testimony of your approbation.

The subject of the following Lecture, which I have the honour of dedicating to you, is of very extensive application. The definition predicates that "its foot rests on earth while its top extends to heaven;" and it embraces all the intermediate steps by which the human soul mounts to immortality. It is an emblem for all time, and embraces interests which no region of the world can change nor any age decay.

When Moses was at the Burning Bush, he was commanded, as an act of reverence, to take off his shoes. And when the Prophet Ezekiel was forbidden to mourn for the loss of his wife, one of the indications of this extraordinary instance was, to "put on his shoes." The present Jews in Barbary, according to Addison, when a death occurs in their family, do not stir abroad for seven days after the interment; or if they should be compelled by any extraordinary or urgent cause to leave their dwelling, it must be barefooted, as a token of reverence to Him by whom they have been stricken.

How much more ought we to express our humility and reverence, when we stand on ground which has been consecrated by Three Grand Offerings, and bears that Holy Book which is the source of all our hopes and comforts. When Jacob occupied the same situation, he said, "surely the Lord is in this place, and I knew it not. And he was afraid, and said, how dreadful is this place! This is none other but the house of God, and this is the Gate of heaven."

That you, my brethren, may ascend the Theological Ladder with the same devout and holy feelings, and in the end receive the reward of your faith, even the salvation of your souls, is the fervent wish of

W. Sir,

And dear Brethren,

Your faithful Servant and Brother.

GEO. OLIVER, D.D.,

Honorary Member of St. George's Lodge.

SCOPWICK VICARAGE,
December 1, 1849.

Lecture the Seventh.

The three Great Lights which form the basis of the Masonic Ladder explained, with a description of the Ladder and its accompaniments.

> "Tyll that I came unto a ryall Gate,
> Where I sawe stondynge the goodly portres,
> Whyche axed me, from whence I came a late;
> To whome I gan in every thynge expresse
> All myne adventure, chaunce, and busynesse,
> And eke my name; I told her every dell;
> Whan she herde this she lyked me right well."
>
> STEPHEN HAWES.

THE next object which attracts our attention in the Symbol of Glory, is the Holy Bible, which is the great charter of a Christian's faith, and anchor of his hope, as well as one of the Great Lights of Masonry. It forms the Tracing Board of the Great Architect of the Universe; and he has laid down there such glorious plans and moral designs, that were we conversant therein and adherent thereto, it would bring us to a building not made with hands, eternal in the heavens. The Bible is the gift of God to man. It is the consummation of wisdom, goodness, and truth. Many other books are good, but none are so good as this. All other books may be dispensed with; but this is absolutely necessary to our happiness here, and our salvation hereafter. It is the most ancient record of facts known in the world; the materials of its earliest history having been compiled, as is most probable, by Shem, or perhaps by Noah. The Rabbins say that Shem was the instructor of Abraham in the history of former events; and that from Abraham they were naturally transmitted through Isaac, Jacob, and Levi, to Moses. And no injury is done to the just arguments on behalf of the inspiration of Scripture, as

Calmet judiciously observes, if we suppose that Shem wrote the early history of the world; that Abraham wrote family memoirs of what related to himself; that Jacob continued what concerned himself, and that, at length, Moses compiled, arranged, and edited, a copy of the holy works extant in his time. A procedure perfectly analogous to this was conducted by Ezra in a later age; on whose edition of the Holy Scripture our faith now rests, as it rests, in like manner, on the prior edition of Moses, if he were the editor of some parts; or on his authority, if he were the writer of the whole.

The evidences of its truth do not depend on the uncertain deductions of human reason, but upon the teaching of the Holy Spirit of God. Its details are confirmed by signs, and wonders, and manifestations of the divine power. On its veracity our holy religion must stand or fall; and therefore our hopes of salvation anchor upon it, as on a rock which can never give way. It is the pillar and ground of Truth; the pedestal and support of Faith; and hence the Masonic Ladder is planted there as on a foundation that cannot be shaken; because its divine author is Jehovah himself. Wisdom, Strength, and Beauty, centre in its pages; for its wisdom is Faith, its strength is Hope, and its beauty is Charity; a double triad which constitutes Perfection; and realized in the pentalpha, which, in the symbolization equally of Christianity and Masonry, refers to the two natures of the incarnate Deity.

This First Great Light of Masonry is not only perfect, but free from every admixture of imperfection; for if the slightest doubt could be raised respecting the truth of any single fact or doctrine which it contains, it would cease to be the Book of God, and our Faith and Hope would no longer have a solid basis to rest upon. But so long as we believe the Deity to be a wise, and powerful, and perfect being, we must also believe that every thing which emanates from his authority is equally wise, and powerful, and perfect, and consequently worthy of the utmost veneration.

"The events recorded to have happened under the old dispensation are often strikingly prefigurative of those which occur under the new; and the temporal circumstances of the Israelites seem designedly to shadow out the spiritual condition of the Christian church. The connection is ever obvious; and points out the consistency of the Divine purpose, and the harmony deliberately contrived to subsist between both dispensations. Thus in the servitude of Israel are described the sufferings of the church. In the deliverance from Egypt is foreshown its redemption; and the journey through the wilderness is a lively representation of a Christian's pilgrimage through life, to his inheritance in everlasting bliss. So also, without too minute a discussion, it may be observed, that the manna of which the Israelites did eat, and the rock of which they drank, as well as the brazen serpent by which they were healed, were severally typical of correspondent particulars that were to obtain under the Christian establishment; as under the sacrifices and ceremonial service of the church, of which the institution is here recorded, was described the more spiritual worship of the Gospel."[1]

The Bible, as the lectures of Masonry predicate, is the sacred compact from God to man, because he has been pleased to reveal more of his divine will in that Holy Book than by any other means; either by the light of Nature, the aid of Science, or Reason with all her powers. And, therefore, as might be expected, it contains a code of laws and regulations which are adapted to every situation in which a created being can possibly be placed. And it not only incites him to virtue, but furnishes a series of striking examples both of good and evil conduct, that he may avoid the one and practise the other to his eternal advantage.

And further, if it gives copious instructions to rulers and governors that they may perform their exalted duties with strict justice and impartiality, it is no less prolific in its directions to men occupying inferior stations of life, to be obedient to the laws, and to respect the powers under which they live, and by which they are protected. Its precepts extend to the duties of rich and poor, parents and children, husbands and wives, masters and servants. There is not a grade in civil society, from the monarch on his throne, to the peasant between the stilts of a plough, but may find ample instructions for moral go-

[1] Gray's Key, p. 98.

vernment, and the regulation of his desires, in that comprehensive Book. And their universal application and divine origin are manifested by the fact, that those who disregard their operation themselves, display an instinctive respect for every one who professes to take the Bible for a rule of faith, and a guide to the requirements of moral duty.

But this Sacred Volume possesses one peculiar excellence which is denied to every other book. We frequently find an entire code of civil duties embodied in a single passage; which, if universally observed by all classes of society, would turn this earth into a Paradise, and its inhabitants into a band of brothers. If mankind could be persuaded to adopt the rule of mutual assistance and mutual forbearance which is there recommended, and copied in the system of Freemasonry; if they would, on all occasions, *do as they would be done by*, nothing would be wanting to the completion of human happiness.

It was the violation of this rule that made Cain a murderer, and filled the antediluvian world with such violent antipathies and unnatural crimes, as made an universal purgation necessary to cleanse it from its gross pollutions. It was the same disregard to this rule which made Nimrod a hunter of men, Pharaoh an impious contemner of God's judgments; Absalom a rebel, and Judas a thief. And in our own times it arms man against his fellow, and produces all that wickedness and vice which human laws, how stringent soever they may be, have totally failed to banish from the world.

The Book before us contains rules for preserving health by the exercise of Temperance and Chastity; for procuring blessings by the practice of Fidelity, Industry and Zeal; for securing a good reputation by Integrity and a faithful discharge of every trust; and for inheriting the promises by the exercise of Faith, the encouragement of Hope, and the practice of Charity, or the universal love of God and man.

Upon the first Great Light, we find two others—the Square and Compasses; which are varied in their position in every degree, to mark the gradual progress of knowledge; and the former is opened at different passages appropriate to each; for the Bible being considered the rule of a Mason's faith, the Square and Compasses, when united, have the same tendency with respect to his practice. The latter are appropriated to the Grand Master, as the ruler and governor of the Craft, because they are the principal instruments used in the construction of plans, and the formation of ingenious designs; which constitute his especial duty at the erection of magnificent edifices. The former belongs to the whole Craft; because, as they are obligated on it, they are bound to model their actions according to its symbolical directions.

But the peculiar appropriation of the Square is to the Master of a private Lodge. Its utility as an implement of manual labour belonging to operative Masonry, is to try and adjust all irregular corners of buildings, and to assist in bringing rude matter into due form; while to the speculative Mason it conveys a corresponding lesson of duty, teaching him, that by a course of judicious training, the W. M. reduces into due form the rude matter which exists in the mind of a candidate for initiation; and thus, being modelled on the true principles of genuine Masonry, it becomes like the polished corners of the Temple. And by virtue of this jewel, which sparkles on his breast, he is enabled to cause all animosities, if any such should unfortunately exist among the brethren, to subside, that order and good fellowship may be perfect and complete.

In a word, the Square points out the general duties of the Master of a Lodge, which are, to consider himself subordinate to the Grand Master and his officers; to keep a regular communication with the Grand Lodge; to give no countenance to any irregular Lodge, or any person initiated therein; not to initiate a person without a previous knowledge of his character; to respect genuine brethren, discountenance impostors and all who dissent from the original plan of Masonry; and above all to set an example to the Lodge, of regularity, decorum, and propriety of conduct.

The Square reads a lesson not less instructive to the whole fraternity; and enjoins them to regulate their actions by Rule and Line, to harmonize their conduct by the principles of morality and virtue, and mutually to encourage each other in the practice of their masonic duties, by the efficacious influence of good example;

which constitutes an additional illustration of the first Great Light. It is, indeed, a remarkable peculiarity of that Holy Book, that it unites precept so closely with example, as to afford instances of moral and religious conduct which will apply to all mankind, rich or poor, with equal effect. And this is one reason why, in the system of Freemasonry, the Bible is so closely connected with the Square and Compasses. If I were to adduce all the instances contained in the first Great Light, I might refer to almost every page; for we can scarcely open the Book, without finding some great example either of good or evil, which may incite us to the practice of virtue or the hatred of vice. The influence which every man possesses in his own particular sphere is very considerable. Our Grand Master Solomon, when a poor man delivered, by his wisdom, a small city from the army of a very powerful monarch, was led to consider the superiority of wisdom above riches; and concluded that as a wise and good man might be extremely useful to those around him by his example, so might a foolish and wicked man do a great deal of mischief by the same means.

In society example is like leaven to a lump of dough; and its influence is so great as to produce the most favorable or prejudicial effects to the interests of mankind. Each individual observes what others do; and thinks there can be no great harm in copying their example. "I am no worse than my neighbors," is very common language; and such reasoners seem to think that they shall be justified in a breach of the moral law by a reference to the conduct of others. But can such a plea be admissible in a Masons' lodge? Does Freemasonry sanction such an unreasonable argument that the vices of one man will be an excuse for those of another? It should rather appear from the general tenor of the doctrines promulgated in the Lodge, that if any person sets a bad example, it would not only affect the reputation of those who follow it, but it would also increase his own responsibility.

The Master of a Lodge is therefore bound to set his brethren an example of morality and justice, which form the true interpretation of the significant Jewel by which he is distinguished; for such is the nature of our constitation, that as some must of necessity rule and teach, so others of course must learn to submit and obey. Humility in both is an essential duty. And at his installation he solemnly declares that he will "work diligently, live creditably, and act honourably by all men; that he will avoid private piques and quarrels, and guard against intemperance and excess; that he will be cautious in his carriage and behaviour, courteous to his brethren, and faithful to the Lodge; and that he will promote the general good of society, cultivate the social virtues, and propagate a knowledge of the art of Masonry, as far as his influence and ability can extend."

By the Compasses, which are appropriated to the Grand Master, we learn to limit our desires in every station, that, rising to eminence by merit, we may live respected and die regretted. This instrument directs us to regulate our lives and conduct by the rules contained in the first Great Light; and our motto is:

Keep within compass, and you will be sure
To avoid many troubles which others endure.

By the same symbol we are reminded of the impartial and unerring justice of the Most High; who, having in his sacred Tracing Board defined the limits of good and evil, will reward or punish us according as we have obeyed or rejected the divine law. This is an important consideration, and worthy the attention of every initiated Mason; because it involves those peculiar doctrines which are characteristic of the Order—man's personal responsibility, the resurrection, and a future state.

In that awful description of the last Judgment, which is recorded in this sacred Tracing Board, Charity or benevolence to our poorer fellow-creatures is made the test of acceptance or exclusion; and this is an eminent masonic virtue; but Bishop Porteus says, "it is an observation of some importance to be impressed on our minds, that although Charity to our neighbour is a stringent duty, yet it is not the only virtue which we ought to practise; for this makes only one of that large assemblage of virtues which are required to make us perfect. We must therefore collect the terms of our salvation, not from any one passage, but from the whole tenor of the Sacred Writings taken together; and if we judge by this rule, which is the only one that can be safely relied on, we shall find that nothing less than a sincere and lively Faith, producing in us, as far as the infirmity of our nature will permit, universal holiness of life, can ever serve to make our final calling and election sure. But thus much we may collect from that Holy Book, that Charity or love to man is one of the most essential duties of our religion, and that to neglect this virtue must be peculiarly dangerous, and render us unfit to appear at the last day before the tribunal of the Judge."

Such is the teaching of the Great Lights of Masonry; and they therefore constitute an appropriate basis for the foot of the Theological Ladder to rest on, whose principal steps are Faith, Hope, and Charity, and whose summit is the Throne of God.

This Ladder contains staves or rounds innumerable, as the emanations of these three great virtues, with angels ascending and descending thereon. A corresponding symbol among the Jews contains no less than fifty rounds, which they call GATES, and are considered as so many degrees of wisdom, or avenues to the attainment of sublime and mysterious truths. It is incumbent on men that they study the Mysteries which contain this ineffable symbol, before they can receive the influx of divine light. The progress of the candidate in the ascent of this ladder is represented as being exceedingly slow, and obstructed by numerous difficulties; and few there are who arrive at the summit. Moses is said to have passed through only forty-nine; and Joshua was unable to penetrate further than the forty-eighth; but even Solomon, whose wisdom surpassed that of all other men, could never open the fiftieth gate, which leads immediately into heaven, and opens on the Throne of the infinite and omnipotent God whom no man can see and live.[2]

Many of our best divines have entertained an opinion that there are some grounds from analogy to conclude, that a scale of beings exists above us, and another below. And Bishop Hurd says that "the belief is almost universal of such a graduated scale ascending from us to God, though the uppermost round of it may still be at an infinite distance from his Throne. But the direct, indeed

[2] Basnage, p. 189, with Authorities.

the only solid proof of its existence, is the Revealed Word, which speaks of angels and archangels, nay myriads of them, disposed in different ranks, and rising above each other with a wonderful harmony and proportion."

The Masonic Ladder was represented by the artists of the middle ages, in the form of a geometrical staircase; and may be seen in an existing specimen on the triumphal arch of S. Maria Maggiore, at Rome. A symbolical gateway, arched over, is placed at the bottom, another about midway up the ascent, and a third at the top. These are the Gates of heaven, which are expanded to admit all those who have faithfully performed their duty to God, their neighbour, and themselves.

These Gates are mentioned at a very early period of the history of mankind, in connection with the Theological Ladder; for Jacob, to whom the supernal vision was vouchsafed, called it the House of God, and the Gate of heaven.[3] And the same imagery is used by the prophets. Our Grand Master David affords a remarkable instance of the existence of a belief that the mansions of bliss are accessible by means of Gates; and he not only speaks of the Gates of death, through which the soul passes before it is "lifted up;"[4] but rejoices that the Gates of righteousness and the Gate of the Lord are open for the righteous to enter in;[5] and describes the heavenly choir, at the resurrection of Christ, as uniting in the joyful chorus, "Lift up your heads, O ye Gates, and be ye lift up (opened) ye everlasting doors, that the King of Glory may come in."[6] In like manner Job mentions "the Gates of death," and "the doors of the shadow of death;" by which he evidently meant the entrance into Sheol, the world of departed spirits.

The Saviour of mankind describes the way that leads to the realms above as being narrow and of difficult ascent, and the Gates thereof so strait, that few will be able to gain admission.[7] And adds that though the Gates of this city are always open, they are not open for every one to enter in, but are closed against "every thing that defileth, or worketh abomination, or maketh a lie;"[8] but are reserved for those who have faithfully performed

[3] Gen. xxviii., 17. [4] Ps. ix., 13, cvii., 18. [5] Ibid. cxviii., 19, 20.
[6] Ibid. xxiv., 7. [7] Matt. vii., 14. [8] Ibid. xvi., 18.

their duty to God and man. He speaks also of the Gates of hell,[9] which the Christian commentators make to be three in number, and call them Death, the Grave, and Destruction. Death being the first Gate which leads to the realms of eternal misery; it is placed at the end of a broad and well beaten path; the Grave comes next; and Destruction is the final Gate opening into the bottomless pit, which the Jews believed to be in the centre of the earth, under the mountains and waters of Palestine. They appropriated, however, to their Gehenna, three different openings to this place of darkness; the first is in the wilderness, and by that Gate Korah, Dathan, and Abiram descended into hell; the second is in the sea, because it is said that Jonah, who was thrown into the sea, cried to God out of the belly of hell;[10] the third is in Jerusalem, for Isaiah tells us that the fire of the Lord is in Zion, and his furnace in Jerusalem.[11] Under this representation the three Gates are Earth, Water, and Fire.

The same image was used by heathen nations who made their Elysium and Tartarus accessible by the same Gates. Servius, the commentator upon Virgil, says that the Inferni are divided into nine circles, accessible by so many Gates. The first contains the souls of infants; the second the souls of those who, through their simplicity, could not conduct themselves like rational creatures; the third, of those who, through despair, had laid violent hands upon themselves; the fourth, of those who perished through extravagant love; the fifth, the souls of warriors; the sixth, of criminals who had suffered a violent death. Passing through the seventh Gate the souls were subjected to purification; which being completed in the eighth, they passed forward through the ninth, being thoroughly purified, into the Elysian fields.[12]

In the most early ages the heathen imagined that there were certain Gates through which the souls were to pass to their infernal abodes; and from thence, it was, that they used this periphrastical form of speech of going to the Gates of hell, to signify a man's dying. Thus

[9] Rev. xxi., 27.
[10] Jonah ii., 3.
[11] Isai. xxxi., 9.
[12] Montf., vol. v., p. 93

Hezekiah speaks, "I said in the cutting off of my days, that I shall go to the Gates of hell."[13] Which figurative expression in that place is understood simply of death; whereas in the New Testament the Gates of hell signify the powers of darkness. The pagans, however, from whom this mode of speech appears to have been borrowed, understood by the Gates of hell the real entrance into Pluto's dominion. These Gates of hell are frequently found in the monuments of Greece and Rome.[14]

The Persians represented the soul, in its progress to the perfection of a better state of existence, as passing up a tall and steep Ladder, consisting of innumerable steps, and opening by seven Gates into so many stages of happiness. Celsus, as cited by Origen, says on this subject; "the first Gate is of lead; the second of tin; the third of brass; the fourth of iron; the fifth of copper; the sixth of silver; and the seventh of gold. The first they attribute to Saturn, pretending that lead denotes the slowness of that planet's course; the second to Venus, which resembles the softness and splendour of tin; the third, for its solidity and firmness, to Jupiter; the fourth to Mercury, because iron and mercury are applicable to all sorts of work; the fifth, which, by reason of its mixture, is of an unequal nature, to Mars; the sixth to the Moon, and the seventh to the Sun, because gold and silver correspond in colour with these two luminaries." Thus the ascent of the Ladder was graduated and adapted to the mythology of the people, and terminated in a blaze of glory; for the Sun was the supreme deity of the Persians, and next to him the Moon.

The three theological virtues, in the Ladder of Freemasonry, are disposed as the guardians of the principal entrances or Gates, which are closely tyled to the cowan, and the guides through the three stages of a mason's career. These may be likened to the same number of parts in a primitive Basilica or Christian church; viz., 1, the portico for the penitents or unbaptised persons; 2, the nave, or church militant, for the catechumens or those who have been received into the congregation; and 3, the chancel or church triumphant, for the perfect Christian.

[13] Isa. xxxviii.

[14] Montf., vol. v., p. 98.

These three graces of a religious life are thus placed, in conformity with the description of their respective characteristics by St. Paul, 1 Cor. xiii.; and being exclusively attached to Christianity, and admitted into no other religion that ever existed on the face of the earth, leaves the Free and Accepted Masonno alternative but to explain them by the Christian ritual; although they have been explained by a transatlantic Mason as follows. "Faith is the genius of Spring; Hope of Summer; and Charity of Autumn. Faith of Spring, because faith and works must always come together; Hope of Summer, because from that point the Sun looks vertically down upon the seeds which have been committed in faith to the fertilizing bosom of the earth; Charity of Autumn, because then the Sun empties his cornucopia into our desiring laps. Faith is the eastern pillar; Charity the western; and Hope the keystone of this Royal Arch."[15] It will be unnecessary to say that I differ in toto from the above author, in his appropriation of these sublime virtues; and solemnly protest against the principle of making Hope instead of Charity the keystone of the arch.

In the symbol before us we see a female figure seated at the foot of a Ladder, like a dignified matron, under a palm tree, with a dove holding an olive leaf in its mouth, perched on one of the branches, and a lamb at her feet. She bears a Cross in one hand, and a Key in the other. In some of our masonic portraitures, we find Faith designated by a patera or cup; which is, however, a more appropriate symbol of the Roman goddess Fides, who bears no resemblance to our companion of Hope and Charity. This deity, who may be considered the representative of Fidelity, had a temple in the Capitol, and her priests wore white veils; and oaths taken in her name were considered peculiarly binding. She was sometimes represented with a Cup, at others with a basket of fruit and ears of corn. Occasionally she was represented by a turtle dove, on account of its faithfulness to its mate. The most usual symbol, however, was the two right hands joined together in the grasp of friendship.

It is true, we sometimes find, amongst the paintings

[15] Fellows. Masonry, p. 284.

and mosaics of the middle ages, the figure of Faith bearing the Patera; as for instance, on the north basement on the shrine of the blessed Virgin at Florence; but this may be accounted for under the supposition that the artists, being Italian, doubtless took the symbol from the visible attributes of the Roman goddess, without ever reflecting that Christian Faith and the Fides of their fanciful pantheon had not a single quality in common, although the name might suggest a similar appropriation.

In heathen nations a Cup was the insignia of Fides, because it was esteemed oracular; and Julius Serenus has explained the Egyptian method of divining by it. The adept filled it with water, and deposited therein thin plates of gold or silver charged with magical characters. The demon was then invoked by certain prescribed forms of incantation; and the enquiries were answered by the cabalistical hieroglyphics on the plates rising to the surface of the water. Some say that if melted wax were poured into the cup, upon the water, it would arrange itself in the form of letters, and thus give a distinct answer to the proposed enquiries. It was for some such purpose that Dido poured out water from a Patera between the horns of a white cow.

> Ipsa tenens dextrâ Pateram pulcherrima Dido
> Candentis vaccæ media inter cornua fudit.

Sometimes the Patera was used by women for the purposes of divination; and for these reasons it can scarcely be esteemed an appropriate symbol of Christian faith.

In fact its use is explicitly forbidden in the Christian system; for St. Paul calls it "the cup of devils." In the heathen sacrifices, as Macknight informs us, the priests, before they poured the wine upon the victim, tasted it themselves, then carried it to the offerers and to those who came with them, that they also might taste it, as joining in the sacrifice and receiving benefit from it. The cup of devils meant, therefore, the sacrifice offered to the demon or idol, and was therefore expressly condemned.

Amongst the professors of a true religion, the Cup appears to be a more appropriate symbol of Temperance than of Faith; and it was always so considered by Jews as well as Christians. At a Jewish feast, the president used to take a cup of wine into his hand, at the commencement of the ceremony, and after solemnly blessing God for it, and for the mercy which was publicly acknowledged, he drank himself, and then circulated it amongst the guests, who also drank, each in his turn. It is called by David, "the cup of salvation," but Jeremiah terms it, "the cup of God's wrath;" in the former case it was used as an incentive to temperance; but in the latter as a denunciation against ebriety; which always occurred at the Jewish carnival of Purim, where, as in the corresponding ceremonies of the Bacchanalia, the rule was, not to leave off drinking while the topers were capable of distinguishing between the phrases—Blessed be Mordecai! and cursed be Haman! For this reason the Cup was also considered by the Jews as an emblem of the chequered mixture of good and evil by which human life is diversified.

The symbols by which Faith is here designated, possess a more dignified reference. The palm tree has always been considered a symbol of victory, because it is so elastic as to bend under any pressure without breaking asunder, and to regain without difficulty its former erect position when the pressure is removed; thus appearing to be impregnable to all attacks. Hence it was assigned by the early Christians to Faith, because St. John says, "this is the *victory* that overcometh the world, even our Faith."[16] For a similar reason it was esteemed an emblem of the immortality to which Faith leads, because the ancients feigned that this tree never decays. Mariti reports the traditions which exist amongst the Arabs respecting this extraordinary tree. They allege that it will live for hundreds of years; and they had not the vestige of a tradition amongst them that either they or any of their ancestors ever saw a palm tree that died of itself. It bears fruit for ever,[17] and therefore is an appropriate emblem of Faith, which, by its fruits, produces immortality and happiness. It was also a symbol of other Christian virtues, viz., justice, innocence, and a pious and virtuous life.

The above interpretation points out, according to

[16] 1 John v., 4.

[17] Ps. xcii., 14.

Pierius,[18] the reason why Faith is symbolized by a Cross, which is the true palm tree of a Christian; and by means of which the Jew and Gentile will ultimately form one church, and profess one faith, according to that saying of the Redeemer, "if I be lifted up I will draw all men to me." Thus the faithful servants of God were marked in their foreheads with the sign of the Cross to distinguish what they were, and to whom they belonged. Now, among Christians, baptism, being the seal of the covenant between God and man, is therefore, by ancient writers, often called the seal, the sign, the mark and character of the Lord; and it was the practice in early times, as it is at present, to make the sign of the Cross upon the foreheads of the parties baptised. The same sign of the Cross was also made at confirmation; and upon many other occasions the Christians signed themselves with the sign of the Cross in their foreheads, as a token that they were not ashamed of a crucified Master; that on the contrary they gloried in the Cross of Christ, and triumphed in that symbol and representation of it.[19]

The dove is the inhabitant of a pure element which we hope one day to obtain by the exercise of Faith. It was therefore used by the early Christians, and adopted by the Free and Accepted Masons who were employed in the erection of our magnificent cathedrals and churches, as a symbol of this divine quality. Its application in this character was very widely disseminated. On the reverse of a coin of Elagabalus, Faith is represented as a sitting figure, holding a turtle dove in one hand, and an ensign in the other, inscribed Fides Exercitus. The olive figured the peace of mind which the true and faithful Mason enjoys in the contemplation of God's perfections through the medium of the glorious symbol under our notice; and the unity and love which they bear to each other.

The Lamb is the representative of the faithful flock of the Good Shepherd; and hence the use of the lamb-skin in a Lodge, as an emblem of innocence, more ancient than the Golden Fleece or Roman Eagle; more honourable than the Star and Garter, the Thistle and Rose, or

[18] Hieroglyphica, fo. 371. C. Ed. Basil, 1575.
[19] Newton on the Prophecies, Diss. xxiv., Part 1.

any other order under the sun which can be conferred by king, prince, or potentate, except he be a Mason. Indeed, white garments, were always considered as distinguishing marks of favour. They were worn in the courts of Princes; and the garments of priests were generally white. They were an emblem of purity, and are therefore interpreted in the Christian system by "the righteousness of saints."[20]

The Cross is a symbol of the eternal life indicated by a perfect religion; in virtue of which, all who believe shall be enabled to start on their Christian course with a full assurance of Hope. Faith will unlock the Portico of the Church militant, that the Christian soldier may enter and commence his warfare with the three great enemies of his soul; and if he should be victorious in the conflict, and continue faithful unto death, the Captain of his salvation will give him a crown of life.[21]

For this purpose Faith is invested with a Key, as a symbol of power and authority; which is especially referred to in the condemnation which was passed on the public teachers in the law of Moses; who are charged with having taken away the Key of knowledge by which the kingdom of God is opened to mankind, in the multitude of false glosses, superstitious traditions, and heterodox interpretations under which they had buried the pure Word of God. These expounders were designated by a golden Key, as the symbol of their office.[22]

The Cross is in her left hand, and the Key in her right; because the former is always conspicuous, while the latter depends on the hand that contains it. If the candidate perceives the Key in the right hand of Faith, it augurs favourably for his masonic progress; because the right hand was esteemed auspicious, and was supposed to point to the east, whence the benign influences of light and heat, motion and life are disseminated. Thus the heathen aruspices, when they made their observations, always stood with their faces towards the north, so that the right hand might point towards the east.

Faith is placed near the Holy Bible, to show that it is the evidence of things not seen, and a sure confidence in things hoped for. By the doctrines therein contained,

[20] Rev. xix., 8. [21] Ibid. ii., 10. [22] Luke xi., 52.

we are taught to believe in the blessings of redemption; and with his faith thus strengthened, the Christian Mason is enabled to ascend the first step on the road to heaven.

This faith naturally creates a Hope that we may be partakers of the promises contained in the volume which is thus recommended to our notice; and, accordingly, Hope is represented by a female figure resting on an anchor, to symbolize "the anchor of the soul" on which our hopes are founded, and bearing the insignia of power. Hence Hope appropriately occupies the centre of the space between earth and heaven; to intimate that if the faithful brother perseveres in the uniform practice of his moral and social duties, not only to God, but also to his neighbour and himself, he will finally overcome all difficulties. Hope will unlock the second gate, and admit the zealous Mason into the Naos of the Temple, where he is allowed to participate in divine things; and then unveils the glories of the Church triumphant. With such an object in view, he manfully labours to ascend the steep acclivity for its attainment. Hope is to the soul what an anchor is to a ship: a sure and steadfast stay amidst the storms of temptation; which when firmly placed upon the rock of Ages in the Holy of Holies, within the veil, will bear him safely through all his difficulties.

In the Heathen mythology, the figure of Hope is generally represented upon medals, a great number of which are furnished by Montfaucon, as a female crowned with flowers, and resting her right hand upon a pillar, with a bee-hive before her, out of which rise flowers and ears of corn. She sometimes holds, in her left hand, poppies; sometimes lilies, and at others, ears of corn. And most of these symbols have, at one time or another, been introduced into Freemasonry.

When Faith shall be rescinded by beholding its glorious object face to face, and Hope shall be superseded by certainty, Charity will still subsist as the virtue of angels and just men made perfect. Its personation is therefore rightly placed at the summit of the Ladder, where we represent it as a female seated, with an infant on her lap, and two children of unequal ages at her knees. She is also invested with the symbolical Key, and has a circular Jewel suspended from a collar round her neck, on which is inscribed a Heart. At this point the Ladder forms a junction with the highest heavens, and penetrates the regions which lead to the throne of God.

The practice of Charity displays itself in relieving the wants, and comforting the distresses of our brethren in the flesh; and this constitutes the chief boast and glory of our divine science. But this is the least and most inferior part of Charity, and if it consisted in nothing more, it would be difficult to determine why St. Paul should have given it such a decided preference over the other two, by saying, "Now abideth Faith, Hope, and Charity, these three, but the greatest of these is Charity."[23] Bishop Horne says, "Love cannot work ill to his neighbour; it can never injure him in his person, his bed, his property, or his character; it cannot so much as conceive a desire for any thing that belongs to him. But it resteth not content with negatives. It not only worketh him no ill, but it must work for him all the good in its power. Is he hungry? It will give him meat. Is he thirsty? It will give him drink. Is he naked? It will clothe him. Is he sick? It will visit him. Is he sorrowful? It will comfort him. Is he in prison? It will go to him, and, if possible, bring him out. Upon this ground, wars must for ever cease among nations, dissentions of every kind among smaller societies, and the individuals that compose them. All must be peace, because all would be love. And thus would every end of the incarnation be accomplished; good will to men, peace on earth, and to God on high, glory to both."

This divine virtue consists in the love of God and man, which is the only perfect and durable quality we can possess. Prophecies shall fail, tongues shall cease, knowledge shall vanish away; even Faith will become useless when we see God as he is; and Hope will be swallowed up in certainty; but Charity will be the employment of just men for everlasting ages. "This benevolent disposition is made the great characteristic of a Christian, the test of obedience, and the mark by which he is to be distinguished. This love for each other includes the qualities of humility, patience, meekness, and bene-

[23] 1 Cor. xiii., 13.

ficence; without which we must live in perpetual discord; and it is so sublime, so rational, and so beneficial, so wisely calculated to correct the depravity, diminish the wickedness, and abate the miseries of human nature, that did we universally practise it, we should soon be relieved from all the inquietudes arising from our unruly passions, as well as from all the injuries to which we are exposed from the indulgence of the same passions in others."[24]

Thus the exercise of Faith and Hope having terminated in Charity, the Mason who is possessed of this divine quality, in its utmost perfection, may justly be deemed to have attained the summit of his profession; figuratively speaking, an ethereal mansion veiled from mortal eye by the starry firmament; and emblematically depicted in a Mason's lodge by stars, which have an allusion to as many regularly made Masons; without which number no lodge is perfect, nor can any candidate be legally initiated therein.

On the whole, to use the language of a writer of the last century, the Ladder was designed for a type and emblem of the covenant of grace, which was in force from the time of man's apostacy, and began to be put in execution at the incarnation of our Saviour Christ, that only Mediator, who opened an intercourse between earth and heaven. To this mystical meaning of the Ladder, the Redeemer is supposed to allude when he says, "hereafter ye shall see heaven open, and the angels of God ascending and descending upon the Son of man."[25]

[24] Soame Jenyns. View of the Internal Evidences of Christianity.

[25] John i., 51.

LECTURE VIII.

Epistle Dedicatory

TO

BRO. THE EARL OF ABOYNE,	P. G. M.
— JOHN TITTERTON,	W. M.
— WILLIAM PIKE,	S. W.
— JOHN ROYCE, JUN.,	J. W. & SEC.
— REV. T. PEDLEY,	CHAPLAIN.
— J. WEBB,	S. D.
— N. BINEY,	J. D.
— W. STRICKLAND,	P. M. & STEWARD,

Of the St. Peter's Lodge, Peterborough.

My Dear Brethren,

Nothing can be more natural than for a Mason to feel a predilection in favour of the Lodge where he first saw light streaming from the east, to convey a new impetus to his understanding, and to invigorate his reason with the bright rays of Truth, as the beams of the rising sun gild objects in the west with a portion of their gorgeous splendour.

My Alma Mater is the St. Peter's Lodge. There I first imbibed those elements of masonic knowledge which formed the ground-work of all my subsequent studies; and I shall never forget the pleasurable sensations with which I listened to the first instructions I received from Bro. Stevens, who was then the Worshipful Master. In the same lodge my masonic regeneration was completed, for there I received all the three degrees. You will not, therefore, wonder that I entertain lively recollections of a community where I became acquainted with a system which has been a source of no ordinary pleasure and satisfaction, amidst the variegated scenes of a long and eventful life.

It was said of the Egyptian Isis, as I had the pleasure of remarking on a personal visit to the lodge in 1843, and I repeat it here to show that no change has taken place in my filial affection and gratitude to the St. Peter's Lodge;—it was said of the Egyptian Isis, who was the mother of the Spurious Freemasonry, that she was all that was, and is, and shall be; and that no mortal was able to remove the veil that covered her. My masonic Mother has acted towards me a kinder and more maternal part. She removed the veil of darkness and ignorance which blinded my eyes and clouded my understanding; displaying to my delighted view all the charms of her philosophy, her morality, her science; a new world of splendour and surpassing beauty, where Faith, Hope, and Charity, form a gradual ascent to the Grand Lodge above; enlightening the studies of geometrical science by the practice of Temperance, Fortitude, Prudence, and Justice; and cheering the road to heaven by the charms of Brotherly Love, Relief, and Truth.

I am grateful to the Lodge of St. Peter for having conferred on me the title of a Master Mason; a title which, like our glorious badge of innocence, I consider to be more ancient, and more honourable, than any other order under the sun; and I trust I have never disgraced the confidence which was then reposed in me. I have ever considered Freemasonry as the best and kindest gift of heaven to man; subordinate only to our most holy religion. I consider it to be an institution where men of all opinions, and all shades of opinion in religion and politics, may meet as on neutral ground, and exchange the right hand of fellowship; may pursue their mental researches into the region of science and morality, without fearing any collision from hostile opinions to sever the links of harmony and brotherly love by which their hearts are cemented and knit together.

The doctrines which arise out of a consideration of the mysterious Ladder of Freemasonry, are of a character so overwhelming, that the mind with difficulty grasps the mighty subject. Freemasonry defines the three principal staves or rounds, leaving the *innumerable* intermediate ones unnoticed, and applies them to those eminent Theological Virtues which no religion but Christianity considers to be imperative on the worshippers of the T G A O T U.

A disquisition on these sublime graces, as applied to the system of Freemasonry, forms the subject of the following Lecture, which is gratefully inscribed to you by

Worshipful Sir,
And dear Brethren,
Your faithful friend,
And Brother,

GEO. OLIVER, D.D.,
Hon. Member of the St. Peter's Lodge.

SCOPWICK VICARAGE,
January 1, 1850.

Lecture the Eighth.

On the Theological Virtues, and their application to Freemasonry.

"When constant FAITH, and holy HOPE shall die,
One lost in certainty, and one in joy;
Then thou, more happy power, fair CHARITY,
Triumphant sister, greatest of the three,
Thy office and thy nature still the same,
Lasting thy lamp, and unconsumed thy flame,
Shalt still survive ———
Shalt stand before the Host of Heaven confest,
For ever blessing and for ever blest."

PRIOR.

"More ancient than the golden fleece,
More dignified than star
Or garter, is the badge of peace,
Whose ministers we are.
It is the badge of innocence
And friendship's holy flame;
And if you ne'er give that offence,
It ne'er will bring thee shame."

BRO. SNEWING.

OF the Theological Virtues it may be truly said, as we have already predicated of the staves or rounds of the Masonic Ladder, that they are innumerable, although Freemasonry classes them under three principal heads, as the generic parents of them all. I have already observed in a previous lecture, that as these virtues have been introduced into Masonry, it will be impossible to treat on them perspicuously without a reference to the Christian system; although I am inclined to think that those who invented the symbol had an eye to the life of man in its three main divisions, youth, manhood, and old age; or in other words it was considered to be typical of the beginning, middle and end of our existence, prefigured by the three degrees of Masonry. These stages, however, on a careful examination, will be found to correspond with the three great virtues which mark the pilgrim's course from this world to the next.

The ancient philosophers, arguing from the universal progress of generation, increase, and decay, held as a general principle, that all things have a beginning, middle, and end; and that the wise man who has begun well, like the gradual process which converts the rough into a perfect ashlar, will pass his life in acts of piety and virtue, till he receives his reward with God, who is all in all; the beginning, middle, and end of every thing just. This reasoning is of universal obligation, and will be found equally applicable to all religions, as well as to the system of Freemasonry.

1. In the Spurious Freemasonry initiation was thought to convey a spiritual regeneration, somewhat similar to that which takes place at the baptism of an infant, according to the ritual of the Church of England. Hence the first initiation was frequently made at a very early period, which was significantly called "the beginning of life," and water was profusely used as the exterior symbol of the new birth. This was a period of innocence; and the candidate was clothed in white robes as the badge of his acquired purity, because white was considered to be the colour most acceptable to the gods. And before he could be further enlightened in the mysterious doctrines of the orgies, it was necessary that he should prepare himself by penance and mortification, and entertain a steadfast faith in the efficacy of the institution to enable him to lead a life of piety and virtue, that he might be prepared, at the close of his existence, to ascend to Elysium, the sacred abode of the celestial deities.

M. Portal says, in his valuable Essay on Symbolical colours, printed in Weale's Architecture, "Christianity reproduces the doctrines taught in the mysteries. Jesus said, unless a man be born again, he cannot see the kingdom of God. The symbol of regeneration was the rebirth of nature in the spring-time, the vegetation of plants, of trees, and the verdure of the fields. The Messiah, going to execution, consecrated this symbol, as he had already established it by the parable of the sower. Bearing his Cross, he said to those who followed him, if they do these things in a green tree, what shall be done in the dry? The green tree designates the regenerated man, as the dry tree is the image of the profane, dead to spiritual life."

White robes were common to the neophyte in every ancient system of religion throughout the whole habitable globe. Even amongst the Jews a similar practice prevailed. The musicians and singers in the services of the Temple, were clothed in white;[1] as are a similar description of men at the present day in our Cathedral and Collegiate Churches. King Solomon, that Great Master of Masonry in Israel, directed his subjects to clothe themselves in white garments,[2] and to let their actions display a corresponding degree of purity and holiness. White is the symbol of truth, and black is the symbol of error. White reflects all luminous rays, which are an emanation from the Deity; while black is the negation of light, and was attributed to the author of evil. The former being the symbol of Truth, and the latter of falsehood. The book of Genesis, as well as the heathen cosmogonies mention the antagonism of light and darkness. The form of this fable varies according to each nation, but the foundation is everywhere the same;—under the symbol of the creation of the world, or the springing of light out of darkness, it presents the picture of initiation and regeneration.[3]

The beginning of life, or infancy, is still characterized, in every class of society, by white robes or ribbons, to denote the sinless innocence of the new-born babe after baptism has washed away the stains of original sin. And the Divinity has promised that every Christian, who should preserve his purity by overcoming the temptations of the world, shall be rewarded with a white stone as a passport into the regions which lie beyond the cloudy canopy; for in that holy place this colour is particularly distinguished. Those who are admitted are clothed in white raiment, ride on white horses, and are seated on white thrones.[4]

Supported and encouraged by these authorities, the early Christians invested the catechumens with a white

[1] 1 Chron. xv., 27. [2] Eccles. ix., 8.
[3] See Weale's Archit., Part. v., p. 23.
[4] Rev. ii., 17, iii., 5–21, vii., 14, vi., 11, xix., 14, xx., 11.

robe, accompanied by this solemn charge: "Receive the white and undefiled garment, and produce it without spot at the great tribunal, that you may obtain eternal life." At the initiation of a candidate into Masonry, the same ceremony is used to characterize his newly acquired purity, and to display the advantages which are now placed within his reach, if he seek after them with diligence, zeal, and a steady faith in their efficacy. He is invested with a lamb's skin or white leather apron, which is the distinguishing badge of a Mason, more ancient and honourable than any existing order, being the badge of innocence and bond of friendship; and he is strongly exhorted that if he never disgrace that glorious symbol of his profession, it will never disgrace him. And at the conclusion of the ceremonies, "FIDELITY" is particularly recommended to his notice; and he is told that if this be his constant practice throughout the chequered scenes of life, "God will assuredly be with him."

This spirit of unwavering Fidelity, says a talented transatlantic Brother, "never shrinks from the declaration of truth, nor cowardly abandons duty in warning a brother of approaching danger, or labouring with affectionate zeal to reclaim his erring footsteps. It teaches us to walk circumspectly ourselves, and to deal kindly and faithfully with each other under all circumstances in life. If a brother is exposed to temptations, we must succour him, and, if need be, throw around him all the safeguards of moral restraint a benevolent heart can devise. Such fidelity, on the part of masonic brethren, would cure many of the evils, and avert many of the misfortunes incident to the weakness and frailties of human nature. It would dry up many a fountain of sorrow, and wipe off many a reproach cast on this ancient Order of men. Such fidelity and tender regard, such zeal and brotherly love, would be strictly in character with masonic principles, a proper discharge of explicit obligations, and a direct approach to the broad line of duty fixed by the ancient landmarks of the Order."[5]

At the beginning of life youth is carefully instructed in the chief truths of his religion, which are the pillar

[5] Town's Prize Essay.

and ground of his Faith; for if the foundation of this grace be not laid in early life, its existence at a more advanced period will be almost hopeless, as the world is now constituted, because Hope and Charity both spring from it, and they are virtues which ought to distinguish the two following stages of life. The newly initiated Mason is therefore exhorted to strengthen his Faith, which is represented as being the evidence of things not seen, the substance of things hoped for; and by which we have an acknowledgment of a Supreme Being, are justified, accepted, and finally received. This being maintained, and bringing forth its fruits, will turn Faith into a vision, and bring us to that ethereal mansion above, where the just exist in perfect bliss to all eternity; where we shall be for ever happy with God, the Great Architect of the Universe, whose only Son died for us, that we might be justified through Faith in his most precious blood.

This Faith is indicated in the colour of symbolical Masonry, viz., sky blue or hyacinth; which the ancient Christian fathers compared to the qualities of the Salamander, which not only lived in flames, but extinguished fire. The hyacinth, they said, if it be placed in a hot furnace, is unaffected, and even extinguishes it. Thus this colour was considered a symbol of enduring Faith, which triumphs over the ardour of the passions and extinguishes them. Blue Masonry, in like manner, enunciates such excellent moral precepts as the fruits of Faith, that were we strictly adherent thereto, we should be exempted from the ever-burning fire of hell.

There is but one method of producing Hope in manhood, and Charity in old age, and that is, to educate children in the true principles of their Faith, or in other words, of religion and virtue. This was so much regarded in the earliest times, when men were little better than barbarians, that we might almost be tempted to believe it was implanted by Nature in the human breast. Plutarch informs us that the children of the Lacedæmonians were brought up from their infancy in obedience to their parents, and profound reverence for all their superiors in age and authority. They were instructed both by precept and example to honour the hoary head; to rise from their seats when an aged man entered the room where they were assembled; and to stand still and remain silent when they met him in the streets until he had passed by. If any one showed himself refractory to the instructions of his tutors or guardians, or even murmured at their reprehensions, they were severely punished; and it was accounted highly dishonourable in their parents if they did not repeat the correction for the folly and injustice of their complaint.

Thus was Faith and confidence implanted at the beginning of life, by those who even did not know what it meant in its true and legitimate sense, but called it by the name of "trust or assurance;" for they were only half civilized, as is exemplified by what the same author says about their hatred of science. "They looked upon speculative sciences and philosophical studies as so much time misspent; and for this reason they would not suffer the professors of them to reside within the limits of their jurisdiction; because they considered them as subjects which debased the excellency of virtue by vain disputations and empty notions."

The Faith thus inculcated amongst the heathen, was inoperative, and therefore ineffectual to promote any good or valuable purpose. Not so the Faith of a Christian Mason. His religious belief is taught to the youthful Christian in the form of a catechism, which contains an epitome of the terms of salvation; and a Confession of Faith, called the Apostles' Creed, is so firmly fixed in the memory of every individual, as never to be eradicated amidst the varied scenes of the most eventful life. But a true Christian Faith is not like that of the heathen, a mere dead principle of assent, opinion, trust, or assurance, but a lively and unshaken belief in things not seen but hoped for. In the language of Bishop Pearson, the very dust of whose writings has been compared by a competent authority to gold, "when anything propounded to us is neither apparent to our sense, nor evident to our understanding, in and of itself, neither certainly to be collected from any clear and necessary connection with the cause from which it proceedeth, or the effects which it naturally produceth, nor is taken up upon any real arguments, or reference to other acknowledged truths, and yet notwithstanding appeareth to us true, not by a manifestation, but attestation of the truth, and so moveth us to assent, not of itself, but by virtue of the testimony given to it, this is said properly to be credible; and an assent unto this, upon such credibility, is, in the proper notion, Faith or belief."[6]

2. When a youth has completed his education, and his Faith is confirmed by a perfect understanding of the basis on which his hopes are founded, he arrives at manhood, and becomes convinced of the necessity of reducing his knowledge to practice in an intercourse with his fellow-creatures. Faith shows him very clearly that it is by the manner in which he discharges the obligations of duty here that he will be judged hereafter; and that his title to reward will be grounded on his faithful performance of the duties he owes to God, his neighbour, and himself. Such reasoning enlightens his soul with the bright beams of Hope, which

——— spring eternal in the human breast,

and show that wise dispensation of Providence, that

Man never is, but always to be blest.

The heathen nations were fully impressed with the validity of such reasoning; and it is strikingly displayed by the Abbe Barthelemi, in a dialogue between Philocles and Lysis, in his learned work called the Travels of Anacharsis. It is rather lengthy, but will amply repay a serious perusal, as the sentiments are purely masonic, and applicable to our present purpose.

Philocles. What service is most pleasing to God?

Lysis. Purity of heart. His favour is sooner to be obtained by virtue than by offerings.

Philocles. Is this doctrine, which is taught by the philosopher, acknowledged also by the priests?

Lysis. They have caused it to be engraven on the gate of the temple of Epidaurus, *Entrance into these places is permitted only to pure souls.* It is loudly declared in our holy ceremonies; in which, when the priest has said, *Who are those who are here assembled?* the multitude reply, *Good and virtuous people.*

Philocles. Have your prayers for their object the goods of this world?

Lysis. No; I know not but they may be hurtful; and

[6] Pearson on the Creed, Art., i.

I should fear lest the Deity, offended at the indiscretion of my petitions, should grant my request.

Philocles. What, then, do you ask of him?

Lysis. To protect me against my passions; to grant me true beauty, which is that of the soul, and the knowledge and virtue of which I have need; to bestow on me the power to refrain from committing any injustice; and, especially, the courage to endure the injustice of others.

Philocles. What ought we to do to render ourselves agreeable to the Deity?

Lysis. To remember that we are ever in his presence, to undertake nothing without imploring his assistance, to aspire in some degree to resemble him by justice and sanctity, to refer to him all our actions, to fulfil punctually the duties of our condition, and to consider as the first of them all, that of being useful to mankind; for the more good we do, the more we merit to be ranked among the number of his children and friends.

Philocles. May we obtain happiness by observing these precepts?

Lysis. Doubtless; since happiness consists in wisdom, and wisdom in the knowledge of God.

Philocles. But this knowledge must be very imperfect.

Lysis. And therefore we can only enjoy perfect happiness in another life.

In the above quotation we have a copious illustration of the moral duties attendant on the virtue of Hope amongst the followers of Pythagoras; and their practice might be recommended to some who have received the initiatory sacrament of baptism. In the lectures of Masonry, Hope is defined to be the anchor of the soul both sure and steadfast; and it is symbolized by the colour of purple, which was assigned by the ancients to death, as the gates of Elysium, because those who have lived piously, hope to be translated to that blessed region, where they will be happy amidst fields of purple roses. Hence Homer pronounces death to be "a purple glory."

The ancients had so great an esteem for this magnificent colour, that it was especially consecrated to the service of the Deity, and was supposed to be capable of appeasing his wrath. Moses made use of cloths of this colour for the Tabernacle, and for the habiliments of the high priest; and the Babylonians, in like manner, clothed their idols in purple. For some such reason this colour was made symbolical of the virtue of Hope, which Christianity as well as Masonry defines to be the anchor of the soul, because it is both sure and steadfast. The same attributes are applied to the purple colour of the ancients, which was so firm, that no length of time could make it fade. Plutarch tells us, in his life of Alexander, that the conqueror found, amongst the treasures of the kings of Persia, a prodigious quantity of purple stuffs, which had been stored up for 180 years, and yet preserved all their primitive lustre and freshness.

Manhood may be compared to the second degree of Masonry, or, "the middle," in the language of the philosophers, not merely because it is intermediate between the first and third, but because it is practical, and teaches the sciences, which constitute the employment of men in the prime of life, when their minds are vigorous, and their bodies active and capable of enduring fatigue. The second degree also, by the splendid appearances in the Middle Chamber, which are but a glimpse of greater glories in reserve for the successful aspirant, who aims at something beyond the veil, infuses a Hope of participating in that more perfect knowledge which is communicated in the sublime degree.

In a communication from the Grand Lodge of Hamburgh to the Grand Lodge of New York, U. S., dated A.D. 1840, the following passage occurs, which is applicable to the subject in hand. "We have one God and Lord; we all Hope for one heaven. This unites the Mason to every man, and teaches him to overlook many faults in others which might else have wounded his feelings, and preserves the good-will of those who would otherwise have been his opponents. Thus Masonry may be made the means of accomplishing the commands of the Great Architect of the Universe. He who is the best Christian, the most faithful man, will be also the best Mason. So let it be in the profane world and in church relations,—live in brotherhood and peace. Let Freemasons be thus united, and they will stand like an impenetrable phalanx, full of joy and the hope of victory." If, therefore, we be faithful unto death, Hope will present us with a crown of life.

Brotherly love is the virtue of the middle period of life, and constitutes a distinguishing characteristic of a Fellowcraft Mason. In primitive times the great body of the fraternity seldom advanced beyond that degree; and it included many other shining virtues, and amongst the rest Fidelity, which is an essential ingredient in Brotherly Love or Friendship; and without fidelity Hope cannot exist. It would seem, therefore, that human happiness is suspended on this virtue. Wherever it is found to exist in perfection, there we may look for the stamina which cements the social condition of man. Lavater has laid it down as an axiom to "examine what, and how, and where, and when, a man praises or censures; he who always, and everywhere, and, as to essentials, in an uniform manner, censures and blames, is a man that may be depended upon."

In this man we see the perfection of fidelity, and with him we may expect to enjoy uninterrupted friendship, which is superior to all worldly pleasures. A mutual interchange of soul and sentiment will produce unalloyed satisfaction, where the feelings and propensities are unrestricted either by suspicion or doubt, and perfect confidence reigns triumphant. It is the feast of reason and the flow of soul, including comfort in affliction, solace in sickness, and consolation amidst the frowns and persecutions of an ungracious world.

This unalloyed friendship, arising out of fidelity, the offspring of Hope, forms one of the chief recommendations of the masonic system; although it is to be feared that in the world the instances of it are not so numerous as might be wished, if we are to credit the opinion of him who said, that though his acquaintances would fill a cathedral, his friends might be contained in the pulpit. A masonic writer of the last century says that the system of Masonry is established on the comprehensive plan of universality. "Were friendship," he continues, "confined to the spot of our nativity, its operation would be partial, and imply a kind of enmity to other nations. Where the interests of one country interfere with those of another, Nature dictates an adherence to the welfare of our own immediate connections; but, such interference apart, the true Mason is a citizen of the world, and his philanthropy extends to all the human race. Uninfluenced by local prejudices, he knows no preference in virtue but according to its degree, from whatever country or clime it may spring."

A striking illustration of the virtue of Fidelity in a subject towards his prince, is recorded of Bishop Latimer; who having preached what was considered to be an offensive sermon before King Henry VIII., he was commanded to apologize from the pulpit in the king's presence on the following Sunday; and for this purpose he commenced his sermon thus: "Hugh Latimer, dost thou know to whom thou art this day to speak? To the high and mighty monarch, the king's most excellent Majesty, who can take away thy life if thou offendest; therefore, take heed thou speakest not a word that may displease. But then consider well, Hugh, dost thou not know from whence thou comest, and upon whose message thou art sent? Even by the Great and Mighty God, who is always present, and who beholdeth all thy ways, and who is able to cast both body and soul into hell together; therefore, take care that thou deliver thy message faithfully;" and he then proceeded to deliver the self-same sermon which he had preached on the preceding Sunday. After dinner the king commanded the bishop's attendance, and asked him how he dared to conduct himself in that offensive manner? He replied that it was in the honest discharge of his duty both to God and the king; and that he could not have acted otherwise with a quiet conscience. His Majesty, contrary to the expectations of the court, applauded his fidelity, and thanked God that he at least possessed one honest and faithful servant.

If this species of fidelity were carried out in all the relations of private life, the effects would promote the general good. A sincere friend is no flatterer. He will reprove error, as well as applaud virtue; and the one is not more necessary to the best interests of his friend than the other. Admonition is as useful to preserve the health of the mind, as medicine is to restore that of the body; and however it may wound a sensible man's self-love, he will receive it as a tribute of friendship of the greatest value. For this reason, our ancient brethren, when they composed those valuable charges which are above all praise, considered it their duty to admonish every brother, in the regulation of his behaviour at home, and in his own neighbourhood, "to act as becomes a moral and a wise man; particularly, not to let his family, friends, and neighbours, know the concerns of the lodge, &c.; but wisely to consult his own honour, and that of the ancient brotherhood."

Such are the instructions given to the Fellowcraft Mason to stimulate his Hope; and if he models his life and conduct by these and similar precepts contained in the Volume at the foot of the Ladder, he will finally, in the expressive language of Masonry, "overcome all difficulties, and inherit a glorious reward."

3. As we descend into the vale of years, the practice of Brotherly Love or Charity, which began in manhood, is consummated at the prime of life; and accordingly in the third and last degree of Masonry, death and the resurrection are plainly set forth. The mortality and corruption of the body, as well as the immortality of the soul are strikingly symbolized by a coffin, skull, and bones, as emblems of the former, surmounted by a blooming sprig of cassia, to symbolize the latter; and the same awful doctrines are typified in the corn, wine, oil, and salt, which are used at the consecration of our lodges; all appertaining to the third degree of Masonry; and like the "achievements of modern chemistry, facilitate and elevate our idea of that splendid change which may pass on the meanest relics of mortality. We had seen, it is granted, more wondrous transformations in Nature, so early, indeed, and so often, that we forgot to consider and admire them; we know that He, by whom all things were made, must have an energy whereby He is able to subdue all things to himself; but when a human artificer, who confessedly knows nothing of the substance of that matter on which he operates, or of that mind by which he investigates its properties, obtains, by sure processes, a vital fluid (oxygen gas) from a coarse mineral; and inflammable air (hydrogen gas) from water; and shining metals (potassium and sodium) from the ashes of wood or sea-weeds; philosophy thus seems, by her own advances, to cast more and more of practical scorn on her own incredulous question, How are the dead raised up,

and with what body do they come? Shall a frail and puny inquisitor of Nature, whose hand and head must soon return to dust, effect changes thus surprising; and He who created the operative hand, the inquisitive eye, the inventive mind—shall He not show us greater works than these, that we may marvel? Measure the probable excellence of the work by the infinite superiority of the agent, and then conceive now magnificently he is able to verify the prophetic words, It is sown in dishonour, it is raised in glory; it is sown in weakness, it is raised in power."[7]

All the above symbols are calculated to show the uncertainty of life, the certainty of death and judgment, and the necessity of practising Charity, and doing the works of a righteous man that we may have a righteous man's inheritance in the kingdom of heaven.

This virtue is indicated in Masonry by the crimson or rose colour, which was a symbol of regeneration; for the candidate is considered perfectly regenerated as a Mason, when he has been raised to the third degree. M. Portal, who is a competent authority on the symbolization of colours, says, there is a relation between rose colour and Christian baptism which opens the doors of the sanctuary; a relation which is again found in the Latin word *rosa*, derived from *ros*, the dew or rain; the rose tree being the image of the regenerated, while dew is the symbol of regeneration. Horapollo tells us that the Egyptians represented the human sciences by water falling from heaven. Among this nation the sciences were within the temple's precincts, and revealed only to the initiated. In their spurious Freemasonry the rose was a symbol of regeneration and love. The ass of Apuleius recovered the human form by eating crimson roses presented to him by the high priest of Isis. In effect it is only by appropriating to himself the Love and Wisdom of the Deity, signified by red and white, and by their union in the rose, that the regenerated neophyte casts away his brutal passions, and becomes truly a man.

Charity forms the basis of the masonic institution; and I shall not consider myself out of order by telling you

[7] Shepherd, Private Devotion, p. 305

what the Lodge Lectures say of it. The definition was originally extracted from a valuable little book which was in every body's hands sixty years ago, called, "Economy of Human Life," and is very expressive of the virtue it is intended to illustrate. Charity! O how lovely in itself! It is the brightness and greatest ornament of our masonic profession. Benevolence, th ecompanion of heaven-born Charity, is an honour to the heart from which it springs; and is by Masons nourished and cherished. Happy is the man who hath sown in his breast the seeds of benevolence, the produce of which is love and charity; he envieth not his neighbour, he believeth not a tale when reported by a slanderer, he forgiveth the injuries of men, and blotteth them out from his recollection. Then let us remember that we are Masons and men; let us ever be ready to assist the needy if it be in our power to do so; and in the most pressing time of necessity let us not withhold a liberal hand, so shall the most heartfelt pleasure reward our labours, and the produce of love and charity will most assuredly follow.

In working out this beneficent principle, which holds the highest rank amongst the Theological virtues, and in the figurative language of Masonry, "will exalt its professors to an ethereal mansion in the skies," Freemasonry has regard to the three stages of destitution—that of infancy, unavoidable misfortune, and extreme old age. For all of these when proved worthy, relief is at hand. The destitute orphans of deceased brethren are placed in schools where they are clothed, taught, and fed; where they are brought up in the practice of religion and virtue; and when they arrive at the proper age, are placed in situations where their previous training may make them good and worthy members of society.

Here, then, we have an unquestionable proof of the operation of Freemasonry on society in general. The Royal Cumberland School for the orphan children of Freemasons was found to operate so beneficially, even at its first establisment in 1789, and was fraught with such an abundance of unalloyed good, that other public bodies soon found it their interest to imitate so laudable an example. On this model the National and British Schools were formed; as well as the School at St. John's Wood for the orphan children of the Clergy, and many others which have a similar end in view; thus proclaiming the extensive advantages to all orders and descriptions of people, which have resulted from the benignant example of Freemasonry.

Again; our aged brethren who have passed their lives in the practice of masonic principles, and have acquired the approbation of mankind in their several stations; if, in the decline of life, misfortune overtakes them, they are entitled to the provisions of an Asylum, and Annuity Fund, which will afford them the means of subsistence; will contribute to make the closing hours of their pilgrimage a scene of serenity and comfort, and enable them to prepare to meet their God at the approach of that period when the wicked cease from troubling and the weary are at rest.

But the influence of masonic charity does not rest here; for one of the fundamental principles of the Order is, "not to halt in the walk of benevolence while anything remains undone." And therefore we have also a fund of Benevolence, from which relief is extended to those whom calamity may have visited; and whose career of usefulness has been clouded by any casual mischance. These donations have, in numerous instances, proved the means of averting ruin, by removing any temporary difficulty which the inadvertence of others may have thrown in the way. The insolvency, for instance, of any person whose dealings have been extensive, may involve many innocent and industrious families in calamity, and in the absence of some friendly aid, may terminate in their utter ruin. In such cases the fund of Benevolence is at hand, and the recommendation of the Officers of a Lodge will produce immediate relief to ward off the danger, and give time for the energies of a worthy man to expand themselves, until he is able, like the Bruce, in ancient Scottish history, to surmount the difficulty, and replace himself in the position from which the unexpected casualty had threatened to remove him.

Such instances redound highly to the credit of the masonic institution, and force a conviction of its utility on the mind, even of the most sceptical casuist. And they do more than this. They extend the benefits of the Order to society in general, by restoring a confidence in worldly affairs which might otherwise be destroyed; and many families may be benefited by the renewed solvency of one, who, but for the aid thus needfully imparted, would perhaps have been the innocent cause of ruin, or at least mischance to others.

The fund of Benevolence also extends its benefits to the widows of worthy brethren, and enables them, by a timely donation, to wind up satisfactorily the worldly affairs of him they have lost, and to place themselves in some station by which they may provide the necessaries of life for the future. And there are many cases on record where the Grand Lodge has bountifully granted to such interesting objects of benevolence, the sum of £50, or even £100, to avert the evils of poverty and indigence from those who have seen better days, and who consequently are quite unprepared for a sudden change from comparative opulence to absolute want.

Such obvious examples of munificent assistance have stamped Masonry with the seal of universal approbation. Its benefits are known, and its benevolent principles being thus applied to promote the interests of virtue, are estimated by the world as the undoubted emanations of a real love for that benignant religion which teaches Charity and Brotherly Love as the perfection of Faith and Hope. There is no alloy to the pleasure which the dispensation of such benefits produces in the mind; and it may be reflected on at all times, as being well pleasing to the Great Architect of the Universe; because it is an axiom which cannot be refuted, that no life is so acceptable to Him, as that which contributes to the welfare of our fellow creatures.

Here, then, Freemasonry stands unrivalled. No other private institution supports so many charities, or contributes so largely to lighten the evils of life. The great moralist says, "man is a transitory being, and his designs must partake of the imperfections of their author. To confer duration is not always in our power. We must snatch the present moment and employ it well, without too much solicitude for the future, and content ourselves with reflecting that our part is performed. He that waits for an opportunity to do much at once, may breathe out his life in idle wishes, and regret, in the last hour, his useless intentions and barren zeal."

The stream of masonic charity is constant and never-failing. Every class contributes to the fund, and every class derives some benefit from its application. The rich are amply repaid in the satisfaction of mind which it produces; and the poor have their reward at times of the most pressing need. Cast thy bread upon the waters, and it will return to thee in many days.

The distinguishing feature in the distribution of our benevolence is that the instances of it are known to very few, even of our own body, and are never proclaimed to the world. Our alms may be truly said to be done in private, as were those of the secret chamber of the Jews, where money was privately contributed for the relief of the poor. There were two chambers in the sanctuary: one of which, called "the chamber of Secrets," was the place where pious persons deposited their charities for the maintenance of poor children. The Jews appear to have entertained a very high opinion of the merit of private charities. R. Jannai, seeing a certain person give a piece of money to a poor man, told him that it would have been much better to have given him nothing than to have done it so publicly. And our Saviour Christ mentions benevolence before prayer; intimating that it would be well to precede all supplications to God by the practice of charity to man.

It must not be supposed that Freemasonry confines its charities solely to its own body; for, it was justly represented in the Morning Herald some years ago, that the sum of £3000, contributed in India, principally amongst the Freemasons, had been placed in the Bank of Ireland, to the credit of the Mansion House Committee, for the relief of the destitute poor of that country.

These human institutions of a refined benevolence, which spring from Faith and Hope, are lively indications of that Charity which is divine; the spiritual love of God and our brethren in the flesh. This is the sublime virtue which opens the gates of heaven; symbolized, as we have already seen, by the Rose. And hence Charity or Brotherly Love teaches us to conceal the faults and infirmities of our brethren, or to speak of them *under the Rose*, and endeavour to reclaim them from vice to virtue and lead them to the practice of Religion, whose ways are ways of pleasantness, and all her paths are peace.

If our endeavours to produce this auspicious change in the heart and affections be successful, we shall realize the happiness of the angels of heaven, who are represented as rejoicing over a sinner that repenteth. If he suffer from the pestilential breath of calumny and defamation, Charity incites us to undertake his vindication, and restore his good name. Are we injured, we must forgive; if our enemy is placed in our power, we must be merciful; for Freemasonry teaches us to "cultivate brotherly love, the foundation and capestone, the cement and glory of our ancient fraternity; to avoid all wrangling and quarrelling, all slander and backbiting; not to permit others to slander any honest brother, but to defend his character and do him all good offices, as far as is consistent with our honour and safety."

Charity envieth not—we must not either repine at the good fortune of our brother, or rejoice at his calamities; for Freemasonry teaches us to let "the hand and the heart unite in promoting his welfare, both temporal and spiritual, and rejoicing in his prosperity." Charity thinketh no evil—we must not put a bad construction on our brother's words, because it is impossible for us to know the motives by which he is actuated. And in nine cases out of ten, if we presume to judge him by the standard of our own feelings, we shall be guilty of passing an unmerited sentence upon him, and perhaps also upon ourselves. On the contrary, it is our duty, as Masons, to follow the advice contained in the First Great Light: "love your enemies, bless them that curse you, do good to them that hate you, and pray for them which despitefully use you and persecute you, that you may be the children of your Father which is in heaven."[8]

Freemasonry teaches us further, in the exercise of this universal Charity, to "respect a genuine brother, and if he is in want to relieve him, or direct him how he may be relieved. We must employ him in some good work, or else recommend him to be so employed. But we are not charged to do beyond our ability; only to prefer a poor brother that is a good man and true, before any other poor people in the same circumstances."[9]

This state of perfection the heathen could neither

[8] Matt. v., 44.

[9] Ancient Charges, vi.

understand nor acquire. It is true they pretended that their mysteries would restore the soul to its primitive purity, and release it from those shackles by which it is restricted in its worldly tabernacle; that initiation is the precursor of a happy life here, and an introduction to Elysium hereafter, by the communication of divine knowledge, and a spiritual regeneration. To attain which the neophyte underwent four proofs of purification by the elements. The *earth* represented the darkness of the profane; *water* or baptism was the emblem of exterior regeneration, by triumphing over temptations; *air* designated divine truth, enlightening the understanding of the candidate, as *fire*, or the Supreme Being opened his heart to love divine. The symbolic proofs were purely exterior; they figured the four material spheres through which he must pass before attaining the three heavens represented on earth by the three degrees of initiation, which confer a spiritual regeneration.[10] And when he had passed to the highest degree, he was introduced to an illuminated apartment, the type of Elysium, where, as Apuleius expresses it, having arrived at the gate of death, and seen the dreary abode of Proserpine, he was relieved by passing through the elements, and beheld at midnight the sun shining with meridian splendour. And more than this, Plato tells us that the regenerated candidate saw celestial beauty in all its dazzling radiance, and joining in the blessed anthem, he was admitted to the beatific vision of heaven, and pronounced perfect. The candidate was then dismissed with a formula which enjoined him to "Watch and abstain from evil."

The above was nothing more than a senseless and incomprehensible hypothesis. The heathen philosophers —even the very best of them—while they exercised their disciples with lessons of virtue, practised in secret every revolting vice. The same cannot be justly predicated of Freemasons; who, to their pure precepts, add a corresponding purity of practice; and having thus passed through the Gates of Faith, Hope, and Charity, are admitted into the Grand Lodge above, where Charity constitutes the great bond of perfection and happiness.

[10] Weale's Architecture, part v., p. 4.

What joy more pure, or worthier of our kind,
    Than when the good, the wise, the pious meet,
    By bond of kindred love, or friendship sweet,
Link'd in a fellowship of heart and mind,
And rivalry of worth! Nor shall they find
    More joy from aught in that celestial seat,
    Save from God's presence, than again to greet
Each other's spirits, there to dwell combined
In brotherhood of love. The golden tie,
    Dissolved, again unites. Ordain'd to train
Earth's tenants for their dwelling in the sky,
    Faith lost in sight, and Hope in joy, shall wane,
Their task fulfill'd; but heaven-born Charity,
    God's greatest gift, shall still in heaven remain.[11]

[11] Bp. Mant's Happiness of the Blessed, p. 90.

LECTURE IX.

Epistle Dedicatory

TO

BRO. WILLIAM MOSELY TAYLER,	W. M.
— FRANCIS ELKINGTON,	S. W.
— JOHN ARNOLD,	J. W.
— BENJAMIN HALL,	TREA.
— CHAS. WM. ELKINGTON,	P. M. & HON. SEC.
— JOSIAH YEOMANS ROBINS,	S. D.
— JOSEPH SIMS,	J. D.
— WILLIAM GILLMAN,	DIR. OF CER.
— JOSEPH FRANCIS TAYLOR,	STEWARDS
— JOHN SIMPSON NEWTON,	

Of the First Lodge of Light, Birmingham.

My Dear Brethren,

What can be more appropriate than to dedicate to the Lodge of Light a professed disquisition on the source of all Light—the Throne and peculiar residence of that great and glorious Being who is Light itself, and in whom there is no darkness at all?

The contents of the Sacred Roll of the Law are our guides and directors in the narrow path which leads to the supernal mansions of Light; and this divine property is there displayed as an universal emblem of every good, while its antagonistic principle of darkness symbolizes every thing evil. Light is represented in that

Holy Volume as a symbol of Joy and pleasure, while calamity and affliction are expressed by the figure of "gross darkness and the shadow of death." And hence, in the ancient systems, these two principles represented the antagonism of life and death in both the spiritual and material states. A Lodge of Light is therefore peculiarly a place of "decent enjoyment," and the abode of those intellectual pleasures which leave no sting behind.

As darkness is frequently put for affliction, so is Light for happiness; which is one step in advance of the above interpretation. The evangelical prophet, animated by the prospect of the bright appearance of the Sun of Righteousness to confer blessings and redemption on mankind, breaks out into an enthusiastic expression of the pleasure he derives from the stupendous contemplation of the birth of Light. "Arise, shine; for thy light is come, and the glory of the Lord is risen upon thee. For behold, the darkness shall cover the earth, and gross darkness the people; but the Lord shall arise upon thee, and his glory shall be seen upon thee. And the Gentiles shall come to thy Light, and Kings to the brightness of thy rising."

Such, and so beneficial, to compare small things witn great, may be the rejoicing of the members of the Lodge of Light, while engaged in the practice of an Order which inculcates Faith, Hope, and Charity, as the potent virtues of their station on earth, and by the faithful use of which they may attain to the glories which appertain to the cloudy canopy at the summit of the Masonic Ladder.

Another definition of Light afforded by the Book which adorns the Pedestal of Wisdom, is "spiritual knowledge." This is frequently symbolized by a burning lamp, as the candlestick by which it is supported represents the Church of God, whether Jewish or Christian, as the vehicle of that knowledge; for the one was but a type of the other; although one of the ancient Fathers says, quis in candelabro, nisi redemptor humani generis designatur? However this may be, spiritual knowledge constitutes the third step in Light on the way to glory. And accordingly St. John, one of the great parallels and patrons of Masonry, interprets Light to signify the Christian dispensation; and speaks of the advent of

Christ as THE BIRTH OF LIGHT. "Light *is* come into the world;" and as a learned Mason of the last century truly remarks—the Light here meant can be no other than that of divine revelation, which brought life and immortality along with it. The Christian dispensation is constantly and uniformly described in Holy Writ under the figure of Light, from the time that the first faint glimmering of it appeared at a distance, till it shone forth in its full lustre and glory. It is of the same use to the spiritual, that the light of the sun is to the natural world. It gives life, health, and vigour to God's new creation; it makes the day of salvation to dawn upon us, it opens to us the prospect of another and a better life, and guides us in the way to glory and felicity.

Happily has your Lodge been designated: may its members ever participate in that happiness, not only in the present world, but also in the blessed regions of Light where felicity is perfect, and uninterrupted Charity will reign for ever and ever.

Such is the sincere wish,
My dear Brethren,
Of your faithful Servant and Brother,
GEO. OLIVER, D.D.,
Honorary Member of the Lodge of Light.

SCOPWICK VICARAGE
February 1, 1850.

Lecture the Ninth.

Explanation of the Cloudy Canopy and its attendant symbols at the Summit of the Ladder.

"The pleasant garden, and the crystal stream,
  The tree of life which bears on every bough
  Fruits fit for joy, or healing; on the brow,
Of glorious gold a living diadem;
The thrones which blaze with many a radiant gem;
  The branching palms, the raiment white as snow;
  Are these the joys that heaven's abodes bestow?
Or may they rather earth-found figures seem
Of heavenly bliss?—To me it matters not
  If I but reach the mark, whate'er the prize
Of God's high calling."

BISHOP MANT.

"An ethereal mansion veiled from mortal eye by the starry firmament."

LECTURES OF MASONRY.

"Aristotle admirably describes the wonder which would seize upon men, supposing them to have lived up to a certain age underground, and to be then brought suddenly into the light. Allowing them to have inhabited subterranean palaces, adorned with sculpture and painting, and every ornament procurable by wealth; admit that they might have heard of the power and majesty of the gods; yet how great would be their emotion should the earth open suddenly, and disclose to them the vast scenes we daily witness! the land—the sea—the sky—the prodigious volumes of the clouds—the power of the winds—the Sun, its magnitude, its splendour, gilding the whole earth, filling the whole heaven! And then, the spectacle presented by the face of night! The whole firmament glittering with stars, the increasing or waning moon.—Seeing all these things, could they doubt that there are gods, or that these are their works?"

TRANSLATED FROM CICERO.

THE Cloudy Canopy. That mysterious veil which shrouds the secrets of the Grand Lodge above from human observation. "In my Father's house are many mansions," said that Holy Being whom we address as

T G A O T U, which constitute the reward of all who endeavour to qualify themselves for such an inheritance by the practice of the Theological Virtues; and they are spacious enough for all who may be found worthy at the great and final trial.

Symbolical Masonry has three degrees; the life of man has three stages; the Ladder has three principal steps; and heaven has the same number of gradations,[1] in the highest of which is the Throne of God. The Ladder before us reaches to the ceiling of the Lodge; which, according to the masonic definition of its altitude, is "as high as the heavens;" and, in the language of the most ancient Lodge Lectures with which we are acquainted, is "a cloudy canopy or the clouds of heaven;" referring, probably, to those passages of Scripture which describe the day of Judgment, "when the Son of Man shall come in the clouds, and all his holy angels with him;"[2] and gather all nations together in the Valley of Jehoshaphat.[3] A cloud was always considered an unequivocal token of God's presence;[4] and coming in clouds, or with the clouds of heaven, was an acknowledged Jewish symbol of majesty and power. The Rainbow was an emblem of God's covenant with mankind after the Flood, and is a continual sign that God will perform its conditions. The Jews also believed that the sun and fire were legitimate symbols of a divine appearance. The day of Judgment was therefore proclaimed by a "mighty angel coming down from heaven, clothed with a cloud; a rainbow upon his head; and his face as it were the sun, and his feet as pillars of fire."[5]

In the course of a few years after our glorious symbol was introduced into Masonry, an addition was made to the explanation of the covering of the Lodge, in the words, "a cloudy (or celestial) canopy, sprinkled with golden stars." About the latter end of the century the definition was altered to—"the beautiful cloud and spangled canopy of heaven;" and it is also said of the Deity in the lectures of that day, that "he has stretched forth the heavens as a canopy, and crowned his temple with

[1] 2 Cor. xii., 2.
[2] Dan. vii., 13. Matt. xxiv., 30.
[3] Joel iii., 2.
[4] Isai. vi., 4. 1 Kings viii., 10.
[5] Rev. x., 1.

stars as with a diadem." Our transatlantic brethren say, "the Lodge has a cloudy canopy, a starry decked heaven, where all good Masons hope at last to arrive by the aid of the Theological Ladder, which Jacob, in his vision, saw ascending from earth to heaven."

The gates of heaven, therefore, are represented in our symbol as being hidden amidst clouds and darkness; because our Grand Master David had described the locality of the Deity in these words. "He bowed the heavens also and came down; and it was dark under his feet. He rode upon the cherubims, and did fly; he came flying upon the wings of the wind. He made darkness his secret place; his pavilion round about him with dark water, and thick clouds to cover him. At the brightness of his presence his clouds removed."[6] At the dedication of the Temple this description was verified, for the cloud of glory removed from the Tabernacle into the Temple, filling the whole house with thick and impenetrable darkness; from which a light subsequently broke forth, which was so intense that the priests were unable to bear its oppressive lustre; whence Solomon exclaimed in his sublime prayer on that occasion; "the Lord said he would dwell in the thick darkness."

The summit of the Ladder passes over all appearance of matter; penetrates the open door,[7] and is lost and amalgamated in a flood of eternal Light where dwells the blessed Trinity, to whom be glory and honour for ever and ever.

O! 'tis a glorious city! passing ken
   Of eye, and stretch of thought! earth's cities glow
   With no such lustre, nor such riches show.
Holiness is its name. Each citizen
Is pure and holy. There with sainted men,
   Purged from the native dross of earth below,
   And spirits whose natures no pollution know,
God dwells, and He who once for man was slain,
The Lamb all spotless. Who a post would hold
   Therein, by him must thitherward be trod
The path of holiness. That chosen fold
   Defilement enters not. And lo, in broad
Letters of light its charter is enroll'd,
   NONE BUT THE PURE IN HEART SHALL SEE THEIR GOD."[8]

[6] Psalm xviii., 9-12. [7] Rev. iv, 1.
[8] Bp. Mant's Happiness of the Blessed, p. 65.

It may be deemed presumptuous to attempt a description of the glories of this holy place, which is hidden from mortal view by the cloudy canopy of the Lodge, because so little can be known of the happiness of heaven. It is described in several parts of our First Great Light, as "a continuing city"[9] containing "many mansions;"[10] and as being of the greatest magnificence. The foundations are said to be composed of precious stones, the walls of jasper, the gates of pearl, the streets and mansions of transparent gold, shining with the brilliancy of painted glass. Here are Golden Candlesticks surrounding the Divine Being, whose eyes are as a flame of fire; who holds in his right hand the Seven Stars; whose countenance shineth like the Sun in its strength, and out of whose mouth proceedeth a sharp sword,[11] "to smite," according to the testimony of the prophet Isaiah,[12] "the earth with the rod of his mouth; and with the breath of his lips to slay the wicked."[13]

Amidst those supernal dwellings is a sea of crystal, with a prismatic arch of coloured light, and four and twenty elders seated round about the Throne of God, which is encompassed with a living circle of eyes, to denote Wisdom, Prudence, and Foresight; clothed in raiment of unsullied whiteness, with crowns of gold upon their heads, to represent the glorified saints who have fought the good fight of Faith. Nor has the blessed region any need of Sun or Moon, because it is enlightened by the glory of the Most High, and the surpassing splendour of Him that sitteth on the throne.

Our gross conceptions are incapable of comprehending the sublimity of the glorious scene. Of this, however, we may be certain, that whoever overcometh the three great enemies of his soul, shall be endowed with an enlarged capacity of comprehension as the eyes of Elisha's servant were opened to see the chariots and horses of fire by which his master was protected from the attack of the Syrian forces. Along with this increased enlightenment, will be given white robes, as the symbol of admission; as the Jewish priests were admitted to their office; the simple form of which was, after ascertaining

[9] Heb. xiii., 4. [10] John xiv., 2.
[11] Rev. i., 14–16. [12] Ibid. xi., 4.
[13] See the Landmarks of Masonry, vol. ii., p. 117.

that they were free from personal defects, to clothe them in white garments, and admit them into the courts of the priests. The saints had also given to them a White Stone with a new Name. And this was the custom in all ancient criminal processes where a verdict of acquittal was pronounced. By the same token the victorious Christian receives the approving sentence of God. They will be placed before the throne of God, "and serve him day and night in his Temple; and He that sitteth on the throne shall dwell among them. They shall hunger no more, neither shall they thirst any more, neither shall the Sun light on them, nor any heat. For the Lamb which is in the midst of the throne shall feed them, and shall lead them unto living fountains of water; and God shall wipe away all tears from their eyes."[14]

This eternal residence is symbolized by a Triangle within the Vesica piscis, containing the Allseeing Eye of Providence, and surrounded with a Rainbow, and the host of heaven praising God and saying: "Blessing, and honour, and glory, and power, be unto him that sitteth upon the Throne, and unto the Lamb for ever and ever."[15] It appears extremely probable that Jacob saw these glories, and the Shekinah at the summit of the Ladder; for our Scriptures affirm that "the Lord stood above it;" but the Targum of Onkelos gives the passage, "Gloria Domina stabat super eâ." And Christ himself promises to his disciples that "they shall see the heavens opened, and the angels of God ascending and descending upon the Son of Man."[16]

The Holy Trinity is represented by the three prismatic colours which bound the celestial circle; which was explained by the Theosophical Masons of the last century, as "the centre of a Cross, signifying the Trinity in a globular Rainbow; wherein the *red* signifies the Father's property in a glance of fire; *yellow*, the Son's lustre and majesty; *blue*, the substantiality; the dusky brown, the kingdom of darkness. On such a Rainbow Christ will sit to judge the world at the last day in the valley of Jehoshaphat; and thus is he undivided everywhere, and in that Man who is born of God is the whole undivided

[14] Rev. vii., 15–17. [15] Ibid. v., 13. [16] John i., 51.

heart of God, the Son of Man sitting in the circle of his life upon the Rainbow at the right hand of God."

The equilateral triangle, according to the theory of Pierius,[17] represents POWER by the first angle, WISDOM by the second, and LOVE by the third; and that their union in ONE produces CHARITY, which is the brightest emanation of the Deity. The same machinery occurs in the doctrines of the Jewish cabalists, who deduce it from a passage in the book of Wisdom;[18] where the Sacred Triangle is recognized under three several denominations, viz., Goodness or LOVE, Light or WISDOM, and Creation or POWER, which they consider to be the names of the three spheres which emanate from the throne of God. In the Sacred Writings the Divine Being is represented as seated on an azure throne, surrounded by a red or fiery sphere, in the centre of a rainbow formed of brilliant prismatic colours;[19] blue being the symbol of Wisdom, green of Power, and red of Love. In the ancient initiations, the three degrees correspond to these celestial spheres; and the symbolic colours, red, blue, and green, indicate fire, air, and earth.

Within the triangle is the Allseeing Eye, to symbolize the Omnipresence of God in his watchful care over his creatures; and the equal distribution of those good things which will contribute to make us happy in this world, and invigorate us with the hope of sharing in the inconceivable blessings of another and a better.

> The universal Parent of all good
> Stream'd from the cloudy canopy a flood
> Of Light, conferring bliss without alloy;
> In coruscations brilliant, clear, and bright,
> To clear the candidate's astonished sight,
> And the oblivious darkness to destroy.
> And lest with wilful blindness he should stray
> In devious tracks of error's mazy way
> Plunging himself anew in sin and shame,
> The ascent to heaven is to his mind presented;
> Faith, Hope, and Charity, are there cemented,
> As illustrations of the Sacred Name.[20]

[17] Hieroglyphica, fo. 291, F. Ed. Basil., A.D. 1575.

[18] Wisd. vii., 26.

[19] Ezek. i., 28. Rev. iv., 3.

[20] From an unpublished Ode on Masonry, by the late Rev S. Oliver, rector of Lambley.

The sacred emblem is placed within the vesica piscis; a glory which usually encircles, in some ancient paintings, the whole body of Christ, shaped like a fish, and suggested by the word *ιχθυς*, acrostically formed from the initial letters of his titles, viz.: *Ιησοῦς Χριστος, Θεοῦ υἱὸς Σωτηρ,* Jesus Christ, the Son of God, the Saviour. It is frequently represented in the gable windows of a church to convey the same allusion; and constitutes the aureole which proceeds from the glorified *body* of Christ, as when flowing from the head only it radiates in a circle.

This magnificent appearance is one of the most ancient symbols of Masonry; and, in the opinion of our best architects, constituted the most ineffable secret of our ancient brethren, and had a decided analogy with all the mysteries professed by the first societies of Masons. It was so essential to all their undertakings that it could not be dispensed with. The Dionysiaca, the Syrian, and Egyptian artists used it as a leading principle of their art, and it constituted a token of recognition amongst the Master Masons and Epopts in the Platonic system, although its origin is confessedly Egyptian, and is found variously diversified in the pyramids, temples, tombs, and obelisks of that remarkable people. The early Christian architects and painters soon found out its preeminent utility, introduced it into the composition of their respective productions, and made it a mysterious emblem of the Saviour of mankind.

The subject is also repeated on the Basilidean gems or Abraxas, in the form of an anchor, the flukes of which constitute one side of the vesica piscis, flanked by a fish; and one of these, in the collection of Capello, contains certain letters that express the name of Jesus. On one side is represented a naked man with a radiated crown, bearing a whip in his right hand, and having a star on each side of his legs; on the other side is the anchor, and certain figures of the constellations. The inscription forms a curious combination of the Greek and Latin languages, and is as follows: EICVYC XPECTVZ ΓABRIE ANANIA AME. In this inscription the name of Jesus Christ refers to the figure of the Sun on the opposite side of the gem; for the Basilideans identified Jesus with the Sun.

In the Byzantine Mosaics this symbol frequently

occurs. Lord Lindsay has adduced several instances of the prevalence of this custom. Thus in a Mosaic of the triumphal arch of St. Mary Maggiore at Rome, the Israelites are represented as stoning Moses and Aaron after the punishment of the people for the rebellion of Korah; when they are protected within a vesica piscis thrown over them by a hand from heaven; and the Saviour, or Jehovah, appears with them within that sacred emblem. In a picture of the transfiguration, in the church built on Mount Sinai by Justinian, "the Saviour, within a vesica piscis, is elevated in the air between Moses and Elias, who stand on distinct rocks or peaks of the mountain; the three Apostles below kneel and hide their faces. The glory of our Saviour darts in rays like the spokes of a wheel, beyond the vesica piscis."[21]

Again in a representation of the death of the Blessed Virgin, Christ appears within a vesica piscis holding her soul in his arms. In a picture of the Last Judgment, the Saviour is seen amongst dark clouds, seated on a rainbow within the vesica piscis. On the back of the Tabernacle of the Virgin at Florence, she is represented as being carried up to heaven by angels, and seated on a throne inclosed in the vesica piscis. In the cathedral church of Ely she is seen within the same holy symbol. And in another Last Judgment in the Campo Santa of Pisa, our Saviour and his mother are seated side by side, each on a rainbow and within a vesica piscis; although Lord Lindsay confesses that this is the only instance within his experience of such a co-equal juxtaposition. The symbol of God's Throne, as enunciated in the Mosaic dispensation, was the Mercy Seat of the Ark of Alliance in the Tabernacle of Moses, and Temple of Solomon; and his footstool was the Ark itself. On this magnificent throne the Shekinah was seated, as a lambent cloud of glory in the form of a vesica piscis, the same which guided the Israelites through a pathless and dangerous wilderness to the Promised Land, on their deliverance from Egyptian bondage; and ultimately took its station in the Sanctum Sanctorum of the Tabernacle and Temple upon the Propitiatory, as the guide and protector of the people of Israel.

[21] Christian Art, vol. i., p. 89.

In the opinion of the Primitive Church, this sublime appearance was the Being who communicated divine promises to the patriarchs; or, in other words, Jehovah or Christ.[22] "The whole of the Ark seems like the triumphant chariot of God moved by angels, set forth by the four beasts who drew the chariots of the eastern kings; whose pomp the poets exalted into heaven in the chariots of their gods. This of the true God is represented as moving by angels in the clouds, not as any fixed throne in itself; the power and providence of God, whose chariot hath wheels with eyes, making all the world its circle; though often it took its way to the Tabernacle and Temple. Why cherubims were added, the cause hath been often intimated; to wit, by reason that the Logos appearing as God's Shekinah, was attended with angels, and especially with cherubims."[23]

The Rainbow is a token of God's mercy and faithfulness, as it was pronounced to be after the Flood; and the foundation of God's covenant with man. Its brightness and splendour, produced by the reflection of the Sun, are proper types of a divine appearance.

Behold yon bright, ethereal bow,
With evanescent beauties glow;
The spacious arch streams through the sky
Deck'd with each tint of Nature's dye,
Refracted sunbeams through the shower,
A humid radiance from it pour;
Whilst colour into colour fades.
With blended lights and softening shades.

ATHENEUM.

In the ancient systems of religion in our own country, the Rainbow constituted an object of importance. In the Prose Edda the following passage occurs. "I must now ask, said Gangler, which is the path leading from earth to heaven? That is a senseless question, replied Har, with a smile of derision. Hast thou not been told

[22] "The second chapter of the Ecclesiastical History of Eusebius is wholly spent in the proof of the pre-existence of Christ. And in that place, as also in his Book of Evangelical Demonstration, he insisteth, amongst many other examples, on that of Abraham, to whom God once showed himself by his Son in the similitude of a man at the oak of Mamre." (Ten. Idol, p. 324.

[23] Ibid., p. 340.

that the gods made a bridge from earth to heaven, and called it Bifrost? Thou must surely have seen it; but perhaps thou callest it the Rainbow. It is of three hues, and is constructed with more art than any other work. But strong though it be, it will be broken to pieces when the sons of Muspell shall ride over it to the great combat."

The vesica piscis, in our Symbol, penetrates the centre of the Rainbow like a keystone, whence our Continental brethren represent Christ as "the Keystone of the Arch." Thus, Bro. Blanchard Powers, in his Prize Address to the Companions of the Royal Arch, says, "the High Priest or divine Logos being the Keystone of the masonic institution, it may justly be considered as a moral and religious Order whose foundation is Charity. Charity is the bond of perfection. Faith, Hope, and Charity, may be considered as the three immovable pillars in the masonic economy. Our faith is strong in God, that he will fulfil all that is written in the law and the prophets. Faith emboldens us to lay hold of his word as a rule and guide through the rugged paths of life. Hope, as an anchor of the soul, fills us with a cheering and lively prospect of a glorious immortality in a future state. Charity teaches us benevolence and philanthropy, to alleviate the distresses of our fellow men, to bind up the broken-hearted, to raise those that are depressed in spirit, to soothe the cares of the suffering widow, and to wipe away the tears from the orphan's weeping eyes. And that our dwellings may be the asylum of the distressed stranger—humanity, friendship, and brotherly love, is the cement which unites Masons of all nations, tongues, countries, and people, into one indissoluble bond of cordial friendship."

The Right Hand is a symbol of power and authority, for Isaiah says, "we are all the work of God's Hand."[24] Job complains that he is suffering under the Hand of the Lord;[25] which, according to the testimony of our Grand Master Solomon, is the author of both good and evil.[26] And in another place a magnificent image is displayed of fiery streams of refulgent light, which are represented as issuing from the right Hand of God,[27] to enlighten the

[24] Isa. lxiv., 8. [25] Ibid. xix., 21. [26] Eccles. ii., 24. [27] Hab. iii., 4.

universe. Here the Right Hand is put for the Most High, who is described as an everlasting Light or shining substance, which supersedes the use of the Sun and Moon.[28] It is therefore introduced into our Symbol to signify God the Father; for he says himself, "O house of Israel, cannot I do with you as a potter? Behold, as the clay is in the potter's hand, so are ye in my Hand."[29] In a word, the Divine power is frequently symbolized in the Jewish writings by the figure of a Hand.[30]

To stretch out the Hand signifies to chastise, to exercise severity or justice.[31] Thus God delivered his people out of Egypt with a stretched out Hand, and an arm lifted up; by performing many wonders, and inflicting many chastisements on the Egyptians. It was also symbolical of mercy. "I have stretched out mine Hand all day long towards an ungrateful and rebellious people.[32] I have called, and ye have refused; I have stretched out my Hand, and no man regarded."[33] Hence. in the symbol before us, the Hand points to the Holy Bible as the foundation of the Theological Ladder, and the only true source on which our Faith and Hope can be securely based.

A superstition connected with the Hand, in Central America, may be interesting to the Free and Accepted Mason, as it was doubtless derived from the most ancient times. Stephens[34] says, "in the course of many years' residence on the frontiers, including various journeyings among the tribes, I have had frequent occasion to remark the use of the Right Hand as a symbol; and it is frequently applied to the naked body after its preparation and decoration for sacred or festive dances. And the fact deserves further consideration, from these preparations being generally made in the arcanum of the secret lodge, or some other private place, and with all the skill of the adept's art. The mode of applying it in these cases is by smearing the hand of the operator with white or coloured clay, and impressing it on the breast, the shoulder, or other part of the body. The idea is thus

[28] Isa. lx., 20. [29] Jer. xviii., 6.
[30] 2 Kings iii., 15. Isa. viii., 11. Ezek. iii., 14; viii., 3, &c.
[31] Ps. lv., 11. [32] Isa. lxv., 2. [33] Prov. i., 24.
[34] Yucatan, vol. ii., p. 474.

conveyed that a secret influence, a charm, a mystical power is given, arising from his sanctity, or his proficiency in the occult arts. This use of the Hand is not confined to a single tribe or people. I have noticed it alike among the Dacotahs, the Winnebagoes, and other western tribes, as among the numerous branches of the red race still located east of the Mississippi river, above the latitude of 42°, who speak dialects of the Algonguin language." Whence the earlier artists showed a wise humility in abstaining from representations of the Deity, and his secret influence, except symbolically, by a human Hand.

For these reasons the Hand has been introduced into our Symbol to designate the First Person in the Holy Trinity, as the beneficent author and dispenser of every blessing we enjoy; whence the open Hand, in all ages, has been considered a significant token of liberality and kind heartedness; and the phrase, "a blessing on the open Hand," has passed into a proverb to denote a generous and noble disposition. The authority by which it is introduced here as an emblem of that august personage, is found in the Old York Lectures, which illustrate the three first steps of the winding staircase, by a reference to "the three persons in the Trinity;" the legitimate symbols of all of whom will be found in the diagram before us.

"The heavenly host is divided, according to our ecclesiastical authorities, into three hierarchies, and each hierarchy into three orders, nine, therefore, in all. To the upper hierarchy belong the Seraphim, Cherubim, and Thrones, dwelling nearest to God and in contemplation rather than action, and to whom appertain, severally and distinctively, perfect love, perfect wisdom, and perfect rest. To the middle hierarchy—the Dominations, Virtues, and Powers, to whom are committed the general government of the universe, the gift of miracles in the cause of God, and the office of resisting and casting out devils. To the lower—the Principalities, Archangels, and Angels, entrusted with the rule and ordinance of nations, of provinces or cities, and of individuals of the human race; every man being attended by two angels, the one evil, persuading him to sin, for the exercise of his faith; the other good, suggesting righteousness and truth and protecting him from the former."[35]

These are the angelic messengers of the Deity who ascend and descend the Theological Ladder, at the command of the Most High, to bear messages and dispensations to the sons of men, and return with a report of commissions faithfully executed; and my authority for introducing them into the symbol will be found in a Tracing Board inserted by Bro. Stephen Jones as a Frontispiece to his Masonic Essayist, published at the beginning of the present century, and before the reunion of ancient and modern Masons. And in the degree of Knights of the Holy Sepulchre the following characteristic hymn refers to these seraphic beings:—

Hush! hush! the heavenly choir,
They cleave the air in bright attire;
See, see, the lute each angel brings,
And hark, divinely thus they sing.
To the power divine all glory be given,
By man upon earth and angels in heaven.

In ancient paintings these cherubic figures are represented as in our Engraving; the bodies being concealed in the thick cloud, and nothing appearing but the heads, and wings by which they are supported; and they rest not night and day saying, "Holy, holy, holy, Lord God Almighty, which was, and is, and is to come! Thou art worthy, O Lord, to receive glory, and honour and power, for thou hast created all things, and for thy pleasure they are and were created."[36] We have here a magnificent picture, which symbolizes the Deity surrounded by his ministering spirits, as the Creator of the Universe. He is seated on a throne attended by his angels, which, though innumerable, will be abundantly increased at that period when the great company of the redeemed shall be introduced into this Grand Lodge, and dwell with the Most High for ever and ever.

This sublime scene, which the Jewish cabalists significantly term LIGHT, and feign that it has three divisions, which they denominate, "the ancient Light, the pure Light, and the purified Light," has never been, and cannot be unveiled to mortal eyes; for the gate of heaven is

[35] Lord Lindsay, Christian Art, vol. i., xxxiii. [36] Rev. iv., 11.

closed, and the interior is invisible till death and the resurrection shall improve our vision, and form our mortal body like the glorious body of Christ. It is, however, described in the modulated language of Scripture. Isaiah, Ezekiel, and St. John affirm that they were favoured with the privilege of seeing Jehovah on his throne; but in the opinion of all our best divines, the holy Being who displayed his glory to them, was "the man of sorrows," and not the Supreme EN SAPH, the first person in the Trinity, because no man can see God the Father and live. The same spirit, says Bishop Horsley, " which displayed this glorious vision to Isaiah, has given the interpretation of it by the Evangelist St. John; who tells us that the august personage who sat upon the throne, called by Isaiah, Jehovah, was Jesus Christ, whose train filled the Temple, and whose glory fills the universe. In that sense he was seen by the Apostles and all the inhabitants of Palestine, when he came down from heaven to redeem us from our sins.[37] When Moses saw the glory of God, it appeared like an inconceivably resplendent brightness, or clothed with light, as the appearance is generally represented.[38] And when Daniel mentions the Ancient of days, he undoubtedly meant the Deity, and described him thus, that no visible figure of him might be conceivable. But he adds, "a fiery stream issued and came forth from before him: thousand thousands ministered unto him, and ten thousand times ten thousand stood before him;[39] and he that sat was to look upon like a jasper and a sardine stone; and there was a Rainbow round about the throne, in sight like unto an emerald."[40]

David says, in reference to the thick clouds which form the canopy of the lodge, " He made darkness his secret place; his pavilion round about him were dark waters, and thick clouds of the skies."[41] Or in other words, his dwelling place was surrounded with clouds of thick and impenetrable darkness. And Solomon adds to the same effect, God dwelleth in the thick darkness,[42] in reference to the appearance on Mount Sinai when he delivered the Law to Moses;[43] and the mountain burned

[37] John i., 14. [38] Ezek. i,. 26. [39] Dan. vii., 10.
[40] Rev. iv., 3. [41] Ps. xviii., 11. [42] 1 Kings viii., 12.
[43] Exod. xxiv., 15.

with fire unto the midst of heaven, with "darkness, clouds, and thick darkness;"[44] because, as Bede conjectures, the power of his majesty is incomprehensible, and all speculation on the subject must necessarily be dark and unsatisfactory. The mountain was altogether covered with a dense cloud, but within, "the presence of the Lord was as a devouring fire;" for so it appeared to the people in the camp at the foot of the mountain; but to Moses and his companions on the summit, it was bright and shining like the serene and spangled canopy of heaven, and had the resemblance of a pure and spotless sapphire stone.

It must not be understood from hence that the darkness is God, because St. John says, "God is Light and in him is no darkness at all;"[45] for in fact, with him the darkness, though it is called the light of the wicked,[46] is as brilliant as the day. The same kind of appearance occurred at the dedication of Solomon's Temple. The whole house was filled with a dense cloud, which caused the most impenetrable darkness, in the midst of which a clear light broke forth, which the priests could not bear to look upon; and they were obliged to withdraw until its intensity was abated. Thus, the Deity, as a light and fire, dwelleth in the midst of darkness, and in the same manner that the light of his true religion shone in the Holy Land amidst the darkness of idolatry that enveloped the rest of the world.

To convert the masonic ladder into a reality, which is the only method we can use for the spiritual benefit of the brethren, we must consider the character of those worthy Masons who pass through this life in a sincere endeavour to surmount the difficulties of the ascent by the assistance of Faith, Hope, and Charity, that they may be admitted to a participation of the glories which surround its summit, when the gate of death has closed upon them, and the earth, like an affectionate mother, has opened her arms to receive the crumbling frame.

It is an eternal truth, and worthy of the serious consideration of every one who has been admitted to the Light, that if they aspire to the consummation presented by the cloudy canopy of the Lodge, they must discharge

[44] Deut. iv., 11. [45] 1 John i., 5. [46] Job xxxviii., 15.

their several duties to God, their neighbour, and themselves, faithfully and conscientiously. They must feed the hungry, clothe the naked, visit the sick, be possessed of a tongue of good report, ever ready to protect the interests of their brethren, and on all points strictly adhere to the holy teaching of Masonry. And the time will assuredly come when they will sincerely wish that they had always performed these important duties.

Let every brother begin, without delay, to lift up his eyes to the bright Morning Star whose rising brings peace and salvation to the faithful and obedient of the human race. And let him live in this world as if he were really desirous of a happy eternity; that when the king of terrors shall come, he may be welcomed without fear or amazement; and introduce him to the everlasting blessedness which surrounds the throne of God.

LECTURE X.

Epistle Dedicatory

TO

BRO.	WILLIAM RODEN, M.D.,	D. P. G. M. & W. M.
—	AUGUSTUS TILDEN,	P. G. SUP. W. & S. W.
—	REV. W. W. DOUGLAS,	J. W.
—	JOHN SIMPSON,	P. P. G R. & P. M.
—	THOMAS MARK,	TREA.
—		SEC.
—	JOSEPH BOYCOT,	P. G. S. & S. D.
—	RICHARD PARKES PUNT,	J. D.
—	JOHN G. ROSENSTEIN, M.D.,	M. C.
—	JOHN BURROWS,	P. G. S. & STEWARD,

Of the Royal Standard Lodge, Kidderminster.

MY DEAR BRETHREN,

It was a saying nearly two thousand years ago,

————— quod mediocrum est
Promittunt medici, tractant fabrilia fabri.

I consider my tools to be the symbols of Freemasonry I have served a long, although I must confess, an agreeable apprenticeship to learn their use and application; and if they read a solemn lesson to man, that the effects of a good and useful life will be a happy reward in the regions of light and glory, their study cannot be reprehensible, or interfere, in the slightest degree, with the moral or religious duties of a Christian.

I have taken the liberty of dedicating the following lecture on the application of the Cloudy Canopy to you, my beloved brethren and associates in the holy cause of Masonry. It points to the most sacred things, and embodies the glory of that Divine Personage whose fiat created the world. When Moses came down from Mount Sinai, which was the temporary summit of the ladder of Jacob, or the gate of heaven, his face shone with such splendour of Light that the Israelites could not steadily look upon him, and he threw over it a veil before he ventured to address the people. Hence in Christian symbolism, the Mosaic dispensation is figured as a female whose eyes are covered with a bandage; and is thus sculptured in the door of the Chapter House at Rochester. By communion with Jehovah, the great lawgiver had acquired a portion of the light of God's countenance; so in the prayers and means of grace under a better dispensation, a new light is kindled in our souls, as the two disciples, when conversing with Christ, felt their hearts burn within them like fire.

This result was symbolized by the descent of the divine Comforter at Pentecost, which was not in a fire attended by the darkness of a cloud, as in the case of the Israelites in the wilderness; but in a bright flame resting on each of the Apostles, and ascending, like so many pyramids divided at the apex into two or more tongues of fire; because the doctrines of revelation became clearer as the designs of Providence were more fully developed. The cloud was a symbol of the Law; but the fire is an emblem of the Gospel. In the former case the nimbus was attached to the head of Moses only; but in the latter it was common to all who were present in the Temple, although in other respects the appearances corresponded with each other. In the former was thunder; in the latter the noise of a mighty wind. There the people saw a flame, and here fiery cloven tongues; there the mountain trembled, and here the place where they were gathered together was moved. The Jews heard the sound of a trumpet, but the Christians were more highly favoured; for they were endowed with the power of speaking all languages.

These celestial manifestations were but a repetition of the appearance of T G A O T U, who always displayed his glory in fire and light; and will come, in like manner, at the last day to judge the quick and dead. May every Free and Accepted Mason be prepared to meet him with confidence and joy.

With grateful and fraternal respects,
Believe me to be,
My dear Brethren,
Your faithful Servant and Brother,
GEO. OLIVER, D.D.,
Honorary Member of the Lodge

SCOPWICK VICARAGE,
March 1, 1850

Lecture the Tenth.

Application of the Cloudy Canopy and its attendant Symbols at the summit of the Ladder to Freemasonry.

"The Mason views yon glittering orbs on high,
Fix'd in the vast o'er-arching Canopy,
And from the Architect benignant draws
His humbler actions, less extensive laws;
Benevolence is hence his darling theme,
His waking monitor, his midnight dream.
His eye sheds pity's dew, his hand is near
To wipe away affliction's starting tear;
The widow smiles; compassion waves its wing;
The prisoner leaps for joy; the orphans sing."

MASONIC PROLOGUE, 1775.

"A Hall she sees standing,
Than the Sun fairer,
With its glittering gold roof
Aloft in Gimli.
All men of worth
Shall there abide,
And bliss enjoy
Through countless ages."

SCANDINAVIAN VOLUSPA.

IN all the transactions of the present world, activity is excited by the hope or prospect of some useful advantage as the reward of our toil. This observation was never more strikingly verified than in the rage which is so universally displayed at the present day for investigations in search of gold amidst the wild regions of California. Whatever we may be induced to undertake, success is the object of our ambition; and the disgrace of a failure is so much dreaded, that we strive to the utmost of our ability to prevent it. No exertion is spared which may contribute to that end. It will follow, then, that if this principle of action is strong enough to enable a person to surmount all the obstacles which may impede his attainment of worldly benefits; it may be applied with an equally reasonable prospect of success to the business of Freemasonry, and to the climbing of the Theological Ladder which leads to the Grand Lodge above. And the reward promised to such exertions is this;—"to him that overcometh will I grant to sit with me on my Throne, even as I also overcame, and am set down with my Father on his Throne." Or in other words, those who are faithful and constant in the discharge of their several duties, shall occupy a conspicuous situation in the Cloudy Canopy that crowns the summit of the Ladder; and be rewarded with everlasting honour and glory.

On this account it is that the practice of moral virtue is strongly recommended in the system of Freemasonry, as one of the requisites to make our course successful; where Faith produces Hope, and Hope leads to Charity. For this purpose a symbolical armour is provided, and described in the Book which constitutes one of the Great Lights of Masonry, as an antidote and protection against the wiles of the devil.[1] Whence the true Mason will see the necessity of fighting the good fight of Faith,[2] if he be desirous of the reward. And to show the comforts of such a course, he has the example of an inspired Apostle of Jesus Christ, who assures him that having fought that good fight by keeping the Faith, he is certain of receiving, as the recompense of his labours, a peaceable crown of righteousness.[3]

But there is another example of still greater importance to the Christian Mason to incite him to the habitual practice of the Theological and Cardinal virtues, that he may have a claim to the same crown—that of the Saviour of mankind; and he not only directs him what to do to obtain it, but also promises that if he shall succeed in overcoming the temptations of the devil, he will give him a White Stone, and in the Stone a new name written, which no man knoweth saving he that receiveth it.[4]

In the catalogue of virtues which Freemasonry enjoins upon her members as essential to the observance of every brother who is desirous of attaining the summit of the Ladder, the most prominent is a steadfast belief in God

[1] Eph. iv., 11.
[2] 1 Tim. vi., 12.
[3] 2 Tim. iv., 7, 8.
[4] Rev. ii., 17.

the Great Architect of heaven and earth. This article of Faith is made imperative on every candidate, for the purpose of preventing the introduction of infidelity and atheism into the Lodge. And accordingly the Deity is represented, as we have seen in a preceding Lecture, by a Circle; and in the symbolism of the mediæval ages, by a human Hand amidst the clouds of heaven, in token of infinite power, and an invitation to ascend to the mansions of blessedness by the masonic Ladder, because the gates of Faith, Hope and Charity form the only medium of access to the throne of grace.

No person can be initiated without a previous acknowledgment of this fundamental article of a Mason's creed; as the following formula—the first ceremony that a candidate is subject to—will show. After the aid of the Almighty Father and Supreme Governor of the Universe has been supplicated that the candidate may dedicate and devote his life to his service, and become a true and faithful brother; he professes that in all cases of difficulty and danger he will put his trust in God; and is then assured that as his faith is so well founded, and his trust so firmly displayed, he may safely follow the guidance and direction founded on the precepts revealed by that great and holy Being, with a firm but humble confidence; for where the name of God is invoked, we trust no danger can ensue.

This will constitute an unanswerable argument to those who would persuade the public to believe that a masons' lodge is a school of infidelity, and capable of producing revolution and ruin to States and Empires, as Barruel, Robinson, and others, have vainly endeavoured to prove. On the contrary, the belief and acknowledgment of God the Creator is intended to act as a stimulus to our observance of social and civil order, and an incentive to the practice of morality and virtue.

This bounteous and munificent Being, as is indicated by our symbol of the open Hand, has bestowed upon every man a valuable talent, and it is at his peril to neglect the improvement of it. It is true some have been endowed more liberally than others, but to whomsoever much is given, from him will much be required. If God has given wisdom, or strength, or genius, or scientific knowledge, it is with the gracious intention that these blessings shall be widely promulgated, that they may operate to the general advantage of society, as a means of disseminating knowledge, and conferring benefits on his creatures. The wisdom of the wise ought to be employed in directing the affairs of others;—strength is for mutual protection; and the beautiful cunning of the expert artizan, or the more refined and intellectual beauties of the poet and philosopher, have been communicated to favoured individuals, to adorn society with the produce of their works. All are expected to contribute to one and the same end by the union of their several excellences, and no talent must be dormant, under the penalty of being rejected from those happy regions of eternal Light which illuminate and adorn the summit of the Ladder, and cast into outer darkness, where is weeping and gnashing of teeth.

It is to promote this salutary purpose that the Lectures of Freemasonry have been modelled on the system of mutual instruction; that by rousing the energies of the apathetic brother, and stimulating him by emulation to exercise the gifts which have been bestowed upon him he may learn to perform his part creditably in the station of life where he has been placed, and be hailed with the triumphant salutation, "Well done, good and faithful servant, enter into the joy of your Lord." So true are the words of the poet.

> Honour and fame from no condition rise,
> Act well your part, there all the honour lies.

Freemasonry, in the whole of its illustrations, treats infidelity as an absurd speculation which can neither be proved nor clearly comprehended. Who can demonstrate that there is no God, in the face of the glorious works of Nature, which proclaim his existence with their multifarious voices? Who can prove that man has no soul, when the very reason and intellect employed in the process, emanate from that ethereal tenant of his mortal body, and rise in judgment against him to demolish his hypothesis? But if, by a series of false reasoning, the infidel should be able to persuade himself that there is neither God, nor soul, nor future responsibility, what benefit dare he hope to derive from it? The anticipations of utter annihilation after death must, of all other reflections, be the most gloomy and forbidding. A poor mortal, suffering under miseries and misfortunes, and struggling against hardships and persecutions upon earth, with no hope in another and a better world, is a condition not to be imagined without fear and trembling.

A renunciation of infidelity is tested and proved by a habit of active religion; for he who holds no communion with the Deity by private prayer or public worship, is little better than a practical infidel, let his profession be whatever it may. The Free and Accepted Mason, at his first admission into a lodge, as we have just seen, acknowleges his "trust in God;" but if that trust be not animated and kept alive by a regular practice of devotional observances, of what avail will such an acknowledgment be? If any brother should ask, what benefit should I derive from these observances? I would answer him by other questions equally significant, viz., what benefit arises from the creation of man; and of what use was the appointment of one day in seven for rest and worship, if the Sabbath be not devoted to these holy purposes? Of what use was the revelation of God's will to man, if the Scriptures be not read, or learned, or inwardly digested? Those who never pray live in a continual doubt of God's existence, and possess no steady belief in the moral government of T G A O T U. And therefore prayer forms the very essence of Freemasonry, and accompanies all its ceremonies. A man might as well at once avow his disbelief in the being of a God, as to entertain a doubt of the efficacy of prayer, which is the sole medium of communication with the Throne of Grace; and if it ascend, as it ought to do, through the gates of Faith, Hope and Charity, an assurance is given in the divine Tracing Board, that having passed through the gates of death, the faithful brother will be introduced into those celestial mansions which form the brilliant canopy of a Masons' lodge.

This is the happy result of prayer and an observance of the divine ordinances of religion; all of which have a place assigned to them in the usual rites of the Order. We open and close our lodges;—initiate, pass, and raise our candidates;—congratulate, acknowledge, receive, and exalt our expert brethren, by solemn prayer. The reading of the Scriptures constitutes a regular portion of our stated formulæ; our lodges are consecrated and dedicated by a series of religious services; and the gracious aid of the Most High is invoked on all our labours. By these observances the lodge becomes holy ground, and the worthy and zealous Mason hopes to imbibe a portion of those sacred emanations which stream from such a source, like the rays of enduring light that surround the Throne of God. Thus will the tempter of mankind, by whom our first parents were betrayed into sin and shame, be shorn of his chief power, because he will find his intended victim under the protection of grace.

I shall not enter, in this place, into a disquisition on the necessity of moral excellence and Charity as an unerring test of Faith and Hope, although Masonry recommends and enforces the virtue described in the Second Table delivered to Moses on the Mount, as the fruits of Faith, which is, indeed, one of its acknowledged symbols; but proceed to illustrate the doctrine by a series of practical arguments, drawn entirely from the system of Freemasonry, which show the uncertainty of our tenure in this life, and the necessity of providing for the enjoyment of a better, which is placed above the Cloudy Canopy, and accessible by means of the masonic Ladder.

This Cloudy Canopy, like the legend of the third degree, points out the mutability of all things here below; and therefore Freemasonry uses it as an inducement to the brethren "so to pass through things temporal as finally not to lose the things that are eternal;" or in other words, that the Free and Accepted Mason, having performed the duties recommended to him in the lodge, and passed through the gate of Faith, may gradually ascend the innumerable steps of the Ladder, by a lively Hope of receiving the promised rewards, till he attains to that universal Charity which rejoiceth in the truth. Then he cannot fail to be admitted into the number of the heavenly hierarchy, amongst those happy souls who are permitted to say, "Worthy is the Lamb that was slain to receive power, and riches, and wisdom, and strength, and honour, and glory, and blessing."[5]

To accomplish this desirable result, the Lectures of

[5] Rev. v., 12.

Masonry give the following judicious advice to the brethren: "As the steps of man tread in the devious and uncertain paths of life, and his days are chequered by good and evil; and as in his passage through this short and precarious stage of existence, prosperity sometimes smiles upon him, while at others he is beset with a multitude of evils;—hence our lodges are furnished with a mosaic flooring, to remind us of the precariousness of our situation; to-day success may crown our labours; while to-morrow we may tread the uneven paths of weakness, temptation, adversity, and death. Since, then, such emblems are continually before our eyes, we are taught to boast of nothing, but to walk uprightly and with humility before God and man, considering there is no station of life on which pride can be securely founded. All men have birth, but some are born to more exalted stations than others; yet, when in the grave all are on a level, death destroying all distinctions. Let every brother, then, consider it his duty to act according to the pure dictates of reason and revelation; cultivating harmony, maintaining charity, and living in unity and brotherly love."

Again: the candidate in one of the degrees is instructed that his admission in a state of helpless indigence was emblematic of the birth of man, who, at his entrance into this mortal existence, is equally helpless, and indebted to others even for the preservation of his life. And it further symbolized the principles of active benevolence for relief and consolation in the hour of affliction. Above all he was taught to bend with humility and resignation before the Great Architect of the Universe; to purify his heart from the operation of passion and prejudice, and to prepare it for the reception of Truth from the precepts of Wisdom, to His glory and the good of his fellow creatures. He is further told that by the second degree of Masonry he was enabled to contemplate the high destination at which he might arrive by the application of his intellectual faculties to the study of heavenly science; and that the secrets of Nature and the principles of moral truth were unveiled, for the purpose of impressing upon him a just estimate of those wondrous faculties with which he is endowed; that he may feel the duty which is thereby imposed upon him of cultivating them with unremitting care and attention, that he may become an useful and happy member of society. When his mind has thus been modelled to virtue and science, the third degree presents him with another great and useful lesson—the knowledge of himself. It prepares him, by contemplation, for the closing hour of existence; and when by means of that contemplation it has conducted him through the chequered scenes of prosperity and adversity incident to this mortal life, it finally instructs him how to die.

I have been thus diffuse in my quotations from the Old Lectures, because the above passages are peculiarly adapted to the subject under discussion. And some lodges in this country, towards the close of the last century, introduced into their lectures the following observations on the certainty of death, which were first made public by Bro. Inwood.

"There is no security from the devouring weapon of death. Without another enemy, this one would fill the world with mourning. The mother forgets all the sorrows of her travail, for joy that a man is born into the world; the father receives the infant with a smile of gratitude to the Giver of all goodness. In a very few days, notwithstanding all the mother's care and the father's solicitude, this innocent babe becomes the victim of death. Again; we see the tear of sorrow moistening the cheek of venerable age, while hanging over the corpse of a beloved son or daughter, snatched from life in the bloom of youth and beauty; we see the strong features of manhood distorted by unaffected grief while standing by the grave of a beloved wife; and we often see the disconsolate widow leading her trembling orphans from their departed father's grave; and, before she could leave the hallowed ground, turn round to heave the farewell sigh, for her sorrows are too great to weep.

"If we see all this, we cannot, then, be ignorant that there is no escape from the piercing arrows of death. The thick walls of the royal palace, with the clay-built cottage of the pauper, are equally pregnable to his darts. Strength or weakness; health or sickness; beauty or deformity; riches or poverty; learning or ignorance; all, in one undistinguished level, fall beneath his mighty arm. Wherever he levels his bow, the mark is certain, the victim falls, the silken cord of life is cut in twain, and the mourners weep about the streets; for the re-union of soul and body, when thus separated, exceeds all human power. Such hath been man in every age of the world; such is man in his present most exalted moments; and such is each of us. To-day perhaps prosperity and joy shine upon our persons, and the persons of our beloved friends, and we only feel the sorrows of another's woe. But to-morrow, nay, perhaps, before this day closes its light, some friendly heart may sigh over our breathless corpse—alas my Brother."

This is very beautiful, and if universally adopted by the Masters of Lodges, could not fail to produce a lasting impression on every Mason's heart, and to make it wiser and better. If we are fully confident that we must soon die, and that after death comes judgment, it seems also to follow as a necessary consequence, that we shall feel it our interest to prepare for the event in such a manner as to produce a favourable sentence when that awful day shall come.

And yet experience convinces us that such a proceeding is not always practised. And why? Not that a thoughtless brother entertains the most distant idea that he shall never die, but because he believes that his lease of life will be extended to an indefinite length, and that there will be ample time to prepare for the approach of the last enemy.

This is the great error of man. Life, with all its uncertainties and vicissitudes, is passed in an unceasing struggle for wealth, or honour, or distinction, or anything but what we possess. The preparation for a state more precious than them all, is swallowed up in the fatal gulph of procrastination; and numbers die as they have lived, sacrificing the blessed hope of everlasting life, in the unextinguished thirst after worldly good. Well might the moral poet say;

> Procrastination is the thief of time,
> Year after year it steals till all are fled,
> And to the mercy of a moment leaves
> The vast concerns of an eternal scene.

It has been seen that Freemasonry endeavours to guard the brethren against this fatal error, by illustrations of a character so decided that they cannot be misunderstood; and at the same time so plain and pointed that they cannot be overlooked. Nor will it be too much to say that the teaching of Masonry on these points has been eminently successful; and there are honourable instances of men, whose indifference to the genial influences of religion has been removed by the gentle admonitions of Freemasonry; and who, from a perfect indifference to all religious restraints, have become zealous and practical Christians in the belief that "if they have Hope only in this world, they would be of all men the most miserable."[6]

Thus Masonry is termed the hand-maiden of religion, because it enforces the practical fruits of Faith, without which all religion is vain. Like Christianity it teaches that of the three Theological Virtues Charity is the best and greatest;—it enjoins the strict observance of the Cardinal Virtues;—it enforces the three great moral duties to God, our neighbour, and ourselves;—it inculcates Brotherly Love, Relief, and Truth, as the principal Point of the masonic system;—it recommends for practice those excellences of character, Secresy, Fidelity, and Obedience; and imprints indelibly upon the mind the sacred dictates of Truth, Honour, and Virtue.

In a word, every moral duty which distinguishes the Christian system, forms a gem in the masonic crown; and being recommended by the practice of the brethren, are diffused throughout society; and the pleasing results are manifested in the harmony which adorns and cements the social system, and produces the abundant and salutary fruits of unity and love in this world, with a confident assurance of happiness in that holy place which is symbolized in the Cloudy Canopy of a Masons' lodge.

What better encouragement can be desired, to induce a brother to discharge, habitually and conscientiously, his duty to God, his neighbour, and himself, as he is directed to do in pursuance of his masonic obligations? He has the promise of reward at that period when death, the grand leveller of all human greatness, has drawn his sable curtain round him; and when the last arrow of this our mortal enemy has been dispatched, and his bow broken

[6] 1 Cor. xv., 19.

by the iron hand of Time. Then when the Angel of the Lord declares that Time shall be no more, he will receive possession of an immortal inheritance in those heavenly mansions veiled from mortal eye by the Cloudy Canopy; for the great I AM, the Grand Master of the whole universe, will invite him to enter into his celestial lodge where peace, order, and harmony shall eternally reign.

In these heavenly places he will inherit all things, and become a polished Pillar in the sacred Temple of the Most High; and from a brother Mason in the lodge on earth he will become a Son of God in the lodge in heaven.[7] To prepare for this dignity it will be necessary to rule and govern the passions, to be obedient to all lawful commands, to keep a tongue of good report, and to practise the general precepts of the masonic Order. The brother who does this will be sure to overcome. But he must keep his lodge closely tyled,[8] and maintain a vigilant watch;[9] because at a day and hour when he thinks not of it, the final *report* will be made.[10]

It is true, the conspiring world offers strong temptations to seduce him from his duty; and unless he exercises the strictest caution, will overcome his virtuous resolutions, as was unfortunately the case with the twelve recanting Fellowcrafts. But the faithful brother will be on his guard against these temptations while ascending the numerous steps of the Ladder which leads to heaven, that he may successfully conquer the difficulties of the ascent. And there are difficulties which cannot be surmounted but by the powerful aid of faith. How many of us may truly say, with that eminent brother St. Paul,—there are times when we are particularly desirous of doing right, but still we wander from the path; when we condemn in others what we practise ourselves; and when the good that we would do, we do not; and the evil we would avoid, that we do.

The reason of all this is easy enough to understand. It proceeds from the temptations of the devil, and made murderers of Akirop, Kurmavil, and Gravelot; whose dreadful fate is held forth as a beacon to warn the con-

[7] Rev. xxi., 7. [8] Matt. vi., 6. [9] 2 Tim. ii., 3.
[10] Matt. xxv., 13.

siderate Master Mason of the evil consequences of listening to suggestions which are expressly forbidden at every stage of his masonic progress.

And there is another extreme which must be carefully avoided by every candidate for the hidden glories which lie beyond the summit of the Ladder. How successful soever his onward progress may be, he must beware of taking his stand on the deceitful ground of an imaginary perfection. If he should be so unfortunate as to suffer himself to be led into this fatal error, he will soon find himself miserably deceived; for Freemasonry, in all its varied disquisitions, will show him that the most perfect man the world ever saw, either thinks or does something every day of his life, which reminds him of the corruption of his nature. The well instructed brother will consider it his duty to go on steadily towards perfection in this life, in the assured hope of attaining it in the next. And it is only to be found in the secret recesses of the Cloudy Canopy.

Now there are some who think that, although they may be abdicted to the practice of every vice which stains and degrades our nature, yet so long as they injure nobody but themselves, they may justly be exempted from any serious violation of the laws of social order. But Freemasonry will teach them another lesson, by showing the consequences of evil example; which, like a contagious atmosphere, contaminates everything that floats upon its surface. If this specious plea be seriously examined, its futility will be plainly manifest. A single illustration will suffice to show, that it is equally at variance with the dictates of truth and reason. We will take an extreme case.

The Atheist will fancy, that his denial of a divine providence is purely personal, and does not affect the community at large. It will be observed, *in limine*, that this man cannot possibly be a Mason, because the Order repudiates Infidelity on the very threshold of the lodge, as we have already seen. He may, indeed, live without seriously injuring his neighbour, if he be not addicted to proselyting. But modern experience proves, that no one can entertain extreme heterodox views on any subject, without using every means in his power to force his tenets on the consciences of others. As witness our

Socialists, Chartists, and Teetotallers, who use every species of agency, both private and public, to disseminate their distorted opinions; and will even renounce their oldest and best friends, if they withhold their assent from the doctrines which they propound.

It follows, therefore, that serious injury is inflicted on individuals, and on society at large, by the agency of any one who professes "freedom of thought" in matters of religion, while he practically denies the being of a God; and entertains the delusive belief, that, even if his opinions on this subject should be erroneous, no one suffers by them but himself. Such an argument, if it were founded in truth, would unhinge the whole frame of civil society; religion would become useless—masonic lodges unnecessary—and the Pedestal, with its sacred furniture, little better than a mockery of T G A O T U.

This, however, is the light in which the enemy of mankind would wish to place morality and religion, for the purpose of obstructing our progress through the consecutive gates of the three principal avenues of the Ladder leading from this world to the next. He influences his agents, the Atheist, the Socialist, and their compeers, to persuade mankind that pleasure is the chief purpose for which man was created; and for that purpose offers them all the kingdoms of the world as the reward of their allegiance. But Freemasonry will arm the worthy brother with the symbolical panoply of the Order; the helmet of salvation, the shield of faith, and the sword of the Spirit; that he may triumphantly resist the insidious persecutions of those who would lead him from the direct *line* of truth, to stray beyond the *circle* of duty. And if he comes out of the battle as a conqueror, he will realize the promise of the Most High, "I will be his God, and he shall be my son."[11]

The precise meaning of this promise involves the subject of the present Lecture. It refers to an asylum provided for the good and worthy Mason in the paradise of God; where he will be clothed in white robes; with the Sacred Name of Jehovah inscribed on his forehead. This region of Light is so resplendent with the glory of God and the Lamb, that it has no need of the Sun or of the

[11] Rev. xxi., 7.

Moon to enlighten it; and none can enter there but they whose names are duly registered in the Book of Life.[12] This happy region is concealed from mortal view, by the cloudy canopy at the summit of the masonic Ladder; being surrounded with clouds and thick darkness to us who are in the flesh, but clear and refulgent to the spirits of just men made perfect; but its glories are accessible to the anxious Mason, by an assiduous endeavour to perform his moral and religious duties.

How bright these glorious spirits shine;
 Whence all their white array?
How came they to the blissful seats
 Of everlasting day?
Lo, these are they from sufferings great,
 Who came from realms of light,
And in the blood of Christ have wash'd
 Those robes which shine so bright.

This reward ought to be an object of some importance to every good and worthy brother, who is desirous of making his profession of Masonry subservient to his best and dearest interests. And this is really the ultimate design of the Order, to those who consider it as a spiritual institution calculated to ennoble the moral character of man. For nothing can tend more effectually to induce holiness here, than the prospect of happiness hereafter. Whoever is desirous of sitting on a throne in heaven, must, as the old Prestonian Charges express it, "study the Sacred Law of God as the unerring standard of truth and justice, and regulate his life and actions by its divine precepts in a strict discharge of the several duties of his station." If he have grace to do this, T G A O T U will be his friend in the present world, and will give him an inheritance in the holy and happy mansions which lie beyond the cloudy canopy, when his allotted period of probation shall be ended.

Will it, then, be considered wise to risk the loss of this happiness for the sake of any worldly good, which, how pleasing soever it may appear, will suddenly vanish away, like the evanescent shadows of the morning sun? Whoever thinks otherwise, must have disregarded equally his masonic obligations, his lessons of initiation, and the

[12] Rev. vii., 9, xiv. 1, xxi., 23, 27.

moral investigations which attend his improved progress in the art, by renouncing all thoughts of Him, in whom he professed to put his trust, and of his moral government of the world. For no Mason could be induced, by any consideration, to neglect the duties, so solemnly undertaken in the name and presence of the Most High, if he really believed Him to possess the power of depriving him in a single instant of life and hope, and excluding him from that blessed abode which is hidden from mortal view in the glorious Symbol before us.

Let every zealous brother, who is desirous of ornamenting the Craft which he professes, seriously consider that every round of the Ladder which he surmounts, will bring him nearer to its summit; that the Hand of God beckons him on, and encourages him to proceed; and that the hosts of heaven rejoice at his successful progress. And if he regularly performs his devotions in public and private, and does his duty in the station of life to which he has been called, he will gradually advance through the open Gates of Faith, Hope, and Charity, till he occupy a throne in heaven, and be rewarded with glory and immortality.

Such is the happiness which is attainable by a steady course in the ascending path of the Theological Virtues. Whoever wishes to share in it, will glorify the Sacred Name of God; will extol Him that rideth upon the heavens by his Name Jah, and rejoice before him.[13] And in addition to this, they will be kind and charitable to each other, and practise all the virtues recommended in the system of Freemasonry. It is, indeed, true, and unfortunately so, that there are many amongst us, who do not possess the power of doing much good to their necessitous fellow creatures; but this is of very little consequence, provided they do all the good they can. It is not the extent of the action, but the feeling of the heart which shows the true Mason. Be merciful after thy power, says the First Great Light; "if thou hast much, give plenteously; if thou hast little, do thy diligence gladly to give of that little; for so thou gatherest to thyself a good reward against the day of necessity."[14]

[13] Ps. lxviii., 4.

[14] Tobit iv., 8.

And so it is of all the duties which a Mason is bound by his O. B. to perform. He is not expected to be charitable beyond his ability; but in all cases, whatever he does, he ought to do it gladly and cheerfully; for a kind and sympathetic word is often of more value than the most profuse pecuniary assistance, if it be rendered with a grudging mind.

But if any brother have reason to believe, that he has not performed his sacred obligations to God and man so strictly as he ought to have done, let him lose no time in endeavouring to repair the evil. If he pray with sincerity and zeal, T G A O T U will vouchsafe his aid in the work of reformation, that he may become a true and faithful brother amongst us; and will endue him with a competency of divine wisdom, that by the aid of the mysteries of Masonry, he may in future display the beauties of godliness; and the answer to his petition will be, "He that keepeth my works unto the end, to him will I give power over the nations; and he shall rule them with a rod of iron. And I will give him the MORNING STAR."[15]

[15] Rev. ii., 26, 27, 28.

LECTURE XI.

Epistle Dedicatory

TO

W. M.*
S. W.
J. W.
P. M.
TREA.
SEC.
S. D.
J. D.
STEWARDS.

Of the Rising Star of Western India.

MY DEAR BRETHREN,

It is quite refreshing to a lover of Masonry like myself, to find that its holy principles are flourishing so extensively in the Eastern part of the globe, where they first originated, and enlightening in an equal ratio both Europeans and natives with the brilliancy of its beams. The Rising Star will, as every good Mason anticipates, be a blessing to ages yet unborn; and, like its type, in the centre of the lodge, will herald a state of universal peace, embodied by your Provincial Grand Master in his new Order of the Olive Branch, which may cement the native and European population into one happy people, as

* The Author has not received the names of the officers of the Lodge, and therefore has no alternative but to leave blanks that they may be filled up with the pen.

children of the same Parent, governed by the same laws, and partners in the same beneficent institutions.

I should, indeed, be insensible to all good and holy feeling, were I to remain unimpressed with the most lively sensations of gratitude to you, my brethren, for your kindness in associating my name with your own, in connection with a lodge, from the existence of which so many beneficial results may be expected to ensue. The flattering manner in which the honour was conferred merits my warmest thanks. Proposed in full lodge by the Provincial Grand Master, Dr. Burnes, whom a great authority truly denominates "the far-shining beacon of the Order in India;" carried by acclamation; and conveyed to me by a distinguished native brother, Manackjee Curtsejee, Esq., in highly complimentary terms; it was ultimately confirmed by a formal diploma, transmitted by the same hand in the following year.

Under these circumstances, a Lecture on the Blazing Star may with great propriety be addressed to the brethren of the Rising Star of Western India, not only as a public expression of gratitude, but also as a tribute of friendship, and a small though inadequate return for the distinguished favours I have received at the hands of so respectable and intelligent a body of men. As the heliacal rising of the canicular Star caused all the inhabitants of Egypt to rejoice in its appearance, as a prelude to those prolific inundations which were a blessing to the land, so may the population of Western India rejoice in the existence of their Rising Star, as the harbinger of moral benefits, more valuable than the produce which the Egyptians derived from the overflowing of their sacred river.

May its glory increase with every succeeding year; and its usefulness exceed the most sanguine anticipations of him who has the honour to subscribe himself,

W. Sir, and dear Brethren,
Your truly obliged and faithful Brother,

GEO. OLIVER, D.D.,
Honorary Member of the Lodge.

Scopwick Vicarage,
April 1, 1850.

Lecture the Eleventh

Enquiry into the true Masonic reference of the Blazing Star.

"A Star, in the hieroglyphical system of the pagan onei ro-critics, denoted a god; and this sense the word doubtless acquired from the universally established doctrine of the Gentiles, that each Star was animated by the soul of a hero-god, who had dwelt incarnate upon earth as a descent or avatar of the creative divinity. Balaam, beholding with open eyes the very person who had appeared to him as the anthropomorphic Angel of Jehovah, and from whom he specially received the communications which he was to make to Balak; beholding (I say) with open eyes this person, as the future victorious offspring of Jacob, he was naturally led, from a full knowledge of his divine character, to describe him prophetically by an hieroglyphic which denoted *a God*. The Star, therefore, forètold by Balaam, is the Lawgiver foretold by Jacob. But the Lawgiver foretold by Jacob is the Man Jehovah. Therefore, the Star foretold by Balaam, is the Man Jehovah also."

FABER.

"I have seen a Blazing Star, or the Shekinah, each of whose beams contained one of the Sacred Names; inclosing the letter G within a circle, and also an equilateral triangle, under which was placed the Ark of the Covenant. The circle denoted is eternity, because it is without beginning and without end; the triangle signified ———; the Blazing Star, the light of Providence pointing out the way of Truth; and the letter G, glory, grandeur, and gomel; all referring to the divine Name and perfections."

LECTURE OF THE DEGREE OF SECRET MASTER.

IT is a remarkable fact, and shows how careful the Deity has always been to preserve a strict uniformity in all his gracious revelations to his creatures, that in every covenant which he condescended to make with man, he always manifested himself by the Star-like appearance of a celestial fire, as a symbol of purity and truth. The Covenant with Adam was made by the Shekinah, or Sacred Fire, in which a deliverer was promised, whose appearance was to be announced by a similar phenomenon. It was Jehovah Elohim, translated the Lord God, "the brightness of the Father's glory, and express image of his person,"[1] who appeared in this holy *cloud of Light* to converse with Adam; and what is denominated by Moses "a flaming sword," when the guilty pair were expelled, was also a vision of the pointed flame which denoted the presence of the Deity, and was repeated to Moses at the Burning Bush, and to the Apostles of Christ at Pentecost.

When the Covenant was renewed with Noah, a similar celestial appearance was manifested in a brilliant semicircle of light charged with prismatic colours; and hence we are told that, when the Jews see the rainbow, they offer up their prayers to God as being faithful to his promise. The heathen had also a tradition of the same nature; whence they believed the rainbow to be a symbol of comfort to mankind. The Greeks denominated it the daughter of Wonder, and a sign to mortal man; and its appearance was considered as a messenger of the gods.

To Abraham, the father of the faithful, the Covenant was again repeated, Jehovah appearing like a splendid and bright fire in the midst of clouds;[2] and at the sacrifice of Isaac on Mount Moriah, when it pleased him to substitute a more agreeable victim, the glory of God broke forth from behind a cloudy canopy like a Blazing Star, and forbade the offering, promising to renew the covenant of blessing, as the reward of his prompt and willing obedience.

In like manner Jehovah appeared to Moses in the Bush, as a flaming fire burning with mild radiance but not consuming; like the divine nature of Christ, symbolized by fire, which burned in his material body, symbolized by the Bush, without injuring the frail substance of his human nature. And a still more signal manifestation was made to Moses on the Mount, when[3] he was permitted to see the hinder part of the glory of God, the forepart being of such exceeding brightness that no man can behold it and live; and therefore, on this occasion, the dazzling lustre of the divine presence was graciously veiled by "a covering cloud."

These remarkable circumstances attending the ap-

[1] Heb. i., 3. [2] Gen. xv., 17. [3] Exod. xxxiii., 23.

pearance of T G A O T U to his favoured creatures, will, in some measure, account for the symbol of a Blazing Star being placed in the centre of our lodges; for it would scarcely have had such a conspicuous situation assigned to it by our ancient brethren, if it had not possessed some very sublime reference. In a primitive Trestle Board of Masonry,[4] the Blazing Star represented BEAUTY, and was called "the glory in the centre," being placed exactly in the middle of the Floor Cloth. In correspondence with this allegorical arrangement, the two pillars of the Porch were symbols of WISDOM and STRENGTH. An alteration was subsequently made by Bro. Dunckerley, under the sanction of the Grand Lodge, by which these three qualities, so necessary to the perfection of any magnificent structure, were assigned to the three chief supporters or pillars of the lodge.

The primitive Blazing Star of Masonry had five points. This was a proper representation of Beauty, as displayed in "a building not made with hands," according to the practice of ancient art, in sculpture, painting, and Mosaics. Lord Lindsay, speaking of a Mosaic of S. Clemente at Rome, executed A. D. 1112, describes it as "a most elaborate and beautiful performance, yielding to none in minuteness of detail and delicacy of sentiment, by a resuscitation of the symbolism of early Christianity, and therefore meriting the most attentive examination. The centre of the composition is occupied by the Tree of Life, the Cross, elevated on the Mount of Paradise and the Church, and reaching to a series of FIVE *concentric rainbow-like semicircles, signifying Heaven, from which the hand of God issues, veiled in clouds*, holding a crown of victory, and also two cords with a heart attached to each, allusive possibly to Hosea xi., 4, or Psalm cxviii., 27. To the right and left, *within the circle*, stands the Paschal Lamb with a glory and other ornaments, all having a tendency to the cross form."[5]

The five points therefore in the masonic Blazing Star are in strict accordance with the primitive symbolization of Christian Masons. And as an exposition of the same principle, the Blazing Star, in one of the ineffable degrees

[4] See Hist. Landmarks, vol. i., p. 133.

[5] Christian Art, vol., i. p. 119.

of Masonry, is made to consist of five points, like a royal crown, in the centre of which appears the initial letter of the Sacred Name. They refer to the five equal lights of Masonry, viz., the Bible, Square, Compasses, Key, and Triangle; and as the Blazing Star is said to enlighten the physical, so the five equal points should enlighten the moral condition of a Master in Israel. They denote the five orders of architecture; the five points of fellowship; the five senses, which constitute the physical perfection of man; and the five zones of the world, all of which are peopled with initiated brothers.

In symbolical Masonry the Blazing Star is considered to be an emblem of Prudence; and our Lectures say: "the Blazing Star, or glory in the centre, refers us to that grand luminary the Sun, which enlightens the earth, and by its genial influence dispenses blessings to mankind." This definition is retained in our present mode of working, with some slight verbal alterations. I entertain considerable doubts of its correctness, for the following reasons:

First, because the Sun constitutes one of our legitimate emblems, and therefore its symbol is superfluous. Secondly, because the Sun was substituted for the Supreme God, and became in that character the great object of worship to all heathen antiquity; as is fully proved by Macrobius,[6] who takes great pains to show that Saturn and Jupiter, Apollo, Mars, and Mercury, with a whole host of other deities, were nothing else but the Sun. And the Egyptians assigned, as one great reason for his worship, that his heat and kindly influence brought their favourite garden gods to maturity.[7] This was also the reason why the Stoics interpreted the *genitalia abscissa* of Saturn to mean the same luminary. And the Sun was so universally worshipped in the time of Julius Cæsar, that some nations who were ignorant of the Roman deities, paid their sole adoration to that idol; for he tells us in his Commentaries,[8] that the Germans worshipped no other gods but those visible intelligences which they believed to be interested in their behalf, viz., the Sun, Moon, and Fire.

[6] Saturnal. l. i.

[7] Lactantius, l. i.

[8] De Bel. Gal., l. 6.

The holy prophets of the Jews sometimes compare Jehovah to the Sun; but it is only because that luminary is the most glorious and resplendent part of the creation; the fountain of light and heat; and the principle of life, health, and fructification to his creatures. And for this reason it was introduced into Masonry. But it is not the only author of the blessings and comforts we enjoy in this world, for we are equally indebted to the elements, earth, air, fire, and water; which are all a means of happiness bestowed upon us by the bounty of an allwise Providence. And lastly, because in the opinion of some of our divines, the Sun is the place of hell, or of punishment for lost souls, and it would not therefore have been designated in Freemasonry by so conspicuous an object as the emblem of Prudence.

In another series of Lectures used in the last century, the Blazing Star is thus defined: "It is placed in the centre, ever to be present to the eye of the Mason, that his heart may be attentive to the dictates, and steadfast in the laws, of Prudence; for prudence is the rule of all virtues; prudence is the path which leads to every degree of propriety; prudence is the channel from whence self-approbation for ever flows; she leads us forth to worthy actions; and, as a Blazing Star, enlightens us through the dreary and darksome paths of life."

That section of the Craft which were known by the name of *ancient* Masons, used the following formula at the latter end of the century: "The Blazing Star or glory in the centre reminds us of that awful period when the Almighty delivered the two tables of stone containing the Ten Commandments to his faithful servant Moses on Mount Sinai, when the rays of his divine glory shone so bright, that none could behold it without fear and trembling. It also reminds us of the omnipresence of the Almighty, overshadowing us with his divine love, and dispensing his blessings amongst us; and by its being placed in the centre, it further reminds us, that wherever we may be assembled together, God is in the midst of us, seeing our actions, and observing the secret intents and movements of our hearts."

The continental definition is, "it is no matter whether the figure of which the Blazing Star forms the centre, be a square, triangle, or circle, it still represents the Sacred

Name of God, as an universal spirit who enlivens our hearts, purifies our reason, increases our knowledge, and makes us wiser and better men."

But the Masons who lived nearer to the great revival, and were cotemporary with the celebrated Bro. Dunckerley, a barrister, with royal blood in his veins, whose authority in Masonry was paramount, and by whose opinions all the measures of the Grand Lodge were regulated, applied this symbol in a sense much more appropriate and sublime. It was said to represent "the Star which led the wise men to Bethlehem, proclaiming to mankind the nativity of the Son of God, and here conducting our spiritual progress to the Author of our redemption." And this application of the symbol is blended with the former by our transatlantic brethren in this definition. "The Blazing Star is emblematical of that prudence which ought to appear conspicuous in the conduct of every Mason; and is more especially commemorative of the Star which appeared in the East to guide the wise men to Bethlehem, and proclaim the birth and the presence of the Son of God."

Now it is observable that the land of Judea, where the angels proclaimed "glory to God, peace on earth, and good will towards men," had been a beacon and a Blazing Star to the rest of the world for 1500 years at the least, or 2000 if the calculation be made from the divine manifestations to Abraham, before the Star which indicated the place where Jesus was found by the Magi made its appearance. The whole world were involved in the darkness of idolatry, and the Spurious Freemasonry reigned triumphantly in its deepest caverns, while the land of Canaan, occupying a central situation, was preserved by the allwise Disposer of events, as a Light shining in a dark place. There the true religion displayed its blessings to those who were inclined to profit by them; and constituted a type of that more effulgent blaze of glory which should penetrate to the remotest corners of the earth, when the Great Light from heaven was manifested which was ordained to enlighten every man in whatever part of the world he may dwell.[9]

The Blazing Star which constituted the essence and

[9] John i., 9.

glory of the typical religion, was the Shekinah tabernacling in the Holy of Holies; while that of the true religion was "the Word that was made flesh and dwelt among us, and we beheld his glory, as of the only begotten of the Father, full of grace and truth."[10] The light of this Day-spring, or glorious Star in the East, has illuminated the world, dispersing the darkness of ignorance, and enlightening the minds of men with the blessed rays of divine truth. "St. John was the Morning Star that preceded the Sun of Righteousness at his rising; an event, the glory of which is due to the tender mercy of God, since towards the production of it man can do no more than he can do towards the causing the natural sun to rise upon the earth. The blessed effects of the Day-spring, which then dawned from on high, and gradually increased more and more unto the perfect day, were the dispersion of ignorance, which is the darkness of the intellectual world; the awakening of men from sin, which is the sleep of the soul; and the conversion and direction of their hearts and inclinations into the way of peace; that is, of reconciliation *to God* by the blood of Christ, *to themselves* by the answer of a conscience cleansed from sin, and *to one another* by mutual love."[11]

Now a Star, in hieroglyphical language, always denoted a God. Thus when Balaam predicted that a Star should arise out of Jacob and a Sceptre out of Israel, he referred to the Lawgiver or Shilo, of whom that patriarch had already spoken. A Star out of Jacob, and a God out of Jacob, would therefore be parallel expressions. And who could that God be who should bear the sceptre of Israel as King of kings and Lord of lords, but the Theocratic King of Israel, Jehovah, the Messiah, or Christ?

On this prophecy Bishop Warburton observes, with his usual acuteness, that it "may possibly in some sense relate to David, but without doubt it belongs principally to Christ. Here the metaphor of a Sceptre was common and popular to denote a ruler like David; but the Star, though, like the other, it signified in the prophetic writings a temporal prince or ruler, yet had a secret and hidden meaning likewise; for a Star in the Egyptian hieroglyphics denoted God. Thus God, in the prophet

[10] Ibid. v., 14. [11] Bp. Horne's Life of John.

Amos, reproving the Israelites for their idolatry on their first coming out of Egypt, says, Have ye offered unto me sacrifices and offerings in the wilderness forty years, O house of Israel? But ye have borne the tabernacle of Moloch and Chiun your images, *the Star of your God* which ye made to yourselves.[12] The Star of your God is here a noble figurative expression to signify the image of your God, for a Star being employed in the hieroglyphics to signify God, it is used here with great elegance to signify the material image of a God; the words *the Star of your God* being only a repetition, so usual in the Hebrew tongue, of the preceding—*Chiun, your images;* and not, as some critics suppose, the same with *your God Star,* sidus Deum vestrum. Hence we conclude that the metaphor here used by Balaam of a Star, was of that abstruse, mysterious kind, and so to be understood; and consequently, that it related only to Christ, the eternal Son of God."[13] The Great Architect of the universe is therefore symbolized in Freemasonry by the Blazing Star, as the herald of our salvation.

Almost every divine appearance, from the creation of the world to the advent of Christ, was attended with this luminous appearance, only with different degrees of brilliancy; and therefore the Star in the East, which was seen by the wise men, would have the same reference. And as a prediction of its announcement had been embodied in the Spurious Freemasonry of all nations, we cannot wonder that, when it appeared, they should follow its direction. And their expectations were not deceived, for it conducted them to the Holy Land, and became stationary over the town of Bethlehem, the very place where the expected Deliverer was to be found.[14] It was the same glory of the Lord which, on the night of the nativity, shone round about the pious shepherds; and being probably of a globular form, it ascended along with the celestial choir of angels; and might hence have been visible in its ascent at the distance of five or six hundred miles, diminished to the size of a star, hovering over the land of Judea.

This appearance must have strongly attracted the notice

[12] Amos v., 25, 26. [13] Div. Leg., Book ii., s. 4

[14] Matt. ii., 9.

and excited the speculations of mankind. And if these Magi, as is extremely probable, were the descendants of Balaam who prophesied of this very Star, and also of the school of Daniel, who foretold the precise time of the coming of Messiah, their journey to Palestine is very naturally accounted for; and it is explained in a masonic degree called the Illustrious Order of the Cross; as is also their adoration of the Divine Child, who was a light to lighten the Gentiles, and a glory to his people Israel; the Day-spring from on high; the bright and morning Star; the Day Star which riseth in our hearts.[15] And at his crucifixion, *the Light being extinguished*, universal darkness overspread the face of the whole earth, and obscured the light of the Sun.

It was, indeed, the universal belief of all nations, that the appearance of a new Star should indicate an avatar of the Deity, who should descend upon earth to teach mankind the Truth, and point out the way to everlasting happiness. And Chalcidius, in his commentary on the Timœus of Plato, says, "When this Star had been seen by some truly wise men amongst the Chaldeans, who were well versed in the contemplation of the heavenly bodies, they made enquiry concerning the birth of the Deity; and when they had found him they paid him the worship and adoration which were due to so great a Being."

But the Blazing Star must not be considered merely as the creature which heralded the appearance of T G A O T U, but the expressive symbol of that great Being himself, who is described, as we have just seen, by the magnificent appellations of the Day-spring or Rising Sun;[16] the Day Star;[17] the Morning Star;[18] and the bright or Blazing Star;[19] This, then, is the supernal reference of the Blazing Star of Masonry; attached to a science, which, like the religion it embodies, is universal, and applicable to all times and seasons, and to every people that ever did or ever will exist on our ephemeral globe of earth.

It was from a similar interpretation of the prophecy of Balaam that the Gnostics and Basilideans erroneously identified Christ with the material Sun, which constituted

[15] Luke ii., 23. i., 78. Rev. xxii., 16. 2 Peter i., 19.
[16] Luke ii., 78. [17] 2 Peter, ut supra. [18] Rev. ii., 28.
[19] Rev. xxii., 16.

their Blazing Star. St. Jerome informs us that they gave to the Almighty the monstrous name of Abraxas, pretending that, from the agreement of the import of the Greek letters composing that word with the number of days in the Sun's course, Abraxas was identified with the Sun or Blazing Star, as the latter was identified with Christ. The heathen entertained the same idea with respect to Mithras, who was also considered as a Mediator between God and man. St. Austin explains the doctrine by saying that Basilides entertained the idea that there were 365 heavens, corresponding with the number of days in the ancient year, and with the name Abraxas or the Sun, which was therefore considered to be holy and worthy of veneration. The notation is thus expressed,

Α	Β	Ρ	Α	Χ	Α	Σ
1	2	100	1	60	1	200=365

The same may be said of the word Mithras or Meithras; and it is well understood that those pseudo-christians worshipped the Sun under these two names, both of which signify that luminary; and it is evident from many of the gems which are yet in existence, and have been copiously illustrated by Montfaucon, that they considered Jesus Christ to be the material Sun.

Thus these heretics mingled truth and falsehood, and produced a system which the Rosicrucians of the middle ages remodelled for cabalistical purposes; and the world is indebted to Freemasonry for the exposure of its pernicious principles, and the restoration of symbolical machinery to the primitive design of promulgating the true Faith, and vindicating the purity of divine revelation.

The final manifestation of the Great Architect of the universe is recorded in the ingenious degree of Knight of the East and West, taken from the book of Revelation. "And I saw heaven opened, and behold a White Horse; and he that sat upon him was called Faithful and True; and in righteousness he doth judge and make war. His eyes were as a flame of fire, (Blazing Star), and on his head were many crowns; and he had a Name written that no man knew but himself. And he was clothed with a vesture dipped in blood; and his name is called the Word of God. And the armies which were in heaven followed him upon white horses, clothed in fine linen, white and clean. And out of his mouth goeth a sharp sword, that with it he should smite the nations; and he should rule them with a rod of iron; and he treadeth the wine press of the fierceness and wrath of Almighty God. And he hath on his vesture and on his thigh a name written, KING OF KINGS AND LORD OF LORDS."[20]

The masonic meaning of the Sun, Moon, and Seven Stars, is familiar to every well instructed brother, and it will therefore require only a few words to make it more distinctly understood. They are placed in our symbol, without the Cloudy Canopy, because in the regions which it conceals from our view, they are perfectly useless, being enlightened by the "Glory of God and the Lamb." And even to ourselves, the genial effects of the sun's rays would be deprived of their most essential properties, if they were not modified by the atmosphere which surrounds our globe. "The eye is indebted to it for all the magnificence of sunrise, the full brightness of its meridian height, the chastened radiance of the gloaming, and the clouds that cradle near the setting sun. But for the atmosphere, the rainbow would want its triumphal arch, and the winds would not send their fleecy messengers on errands round the heavens. The cold ether would not shed its snow-feathers on the earth, nor would drops of dew gather on the flowers. The kindly rain would never fall; hail, storm, nor fog diversify the face of the sky. Our naked globe would turn its tanned, unshadowed forehead to the sun, and one dreary, monotonous blaze of light and heat dazzle and burn up all things."

Were there no atmosphere, the evening sun would in a moment set, and, without warning, plunge the earth in darkness. But the air keeps in her hand a sheaf of rays, and lets them slip but slowly through her fingers; so that the shadows of evening gather by degrees, and the flowers have time to bow their heads, and each creature space to find a place of rest and nestle to repose. In the morning the garish sun would, at one bound, burst from the bosom of night and blaze above the horizon; but the air watches for his coming, and sends at first but one

[20] Rev. xix., 11-16.

little ray to announce his approach, and then another and by and by a handful; and so gently draws aside the curtain of night, and slowly lets the light fall on the face of the sleeping earth, till her eyelids open, and like man, she goeth forth again to her labour until the evening."[21]

In the Apocalypse mention is made of "a woman clothed with the sun, with the moon under her feet, and upon her head a crown of twelve stars."[22] According to our best commentators, the woman was a type of the Church of Christ; she was clothed with the sun, to denote the blessing of light and knowledge which this Church enjoys by the gracious goodness of Jesus Christ the Sun of Righteousness. The moon was placed under her feet to indicate the prostration of the Jewish ceremonial law; and the crown of twelve stars was intended as a symbol to denote that the Christian Church had the advantage of being illuminated by the inspiration of the twelve Apostles.

The worship of the Sun was common to most heathen nations. According to Herodian, the Emperor Aurelian erected a magnificent Temple to this deity, in which he placed statues of the Sun and Bel, which, along with the other precious decorations, were brought from Palmyra. Montfaucon[23] has given an image of the Sun as worshipped by the Romans. It is the bust of a man placed upon an eagle, having his head surrounded with a nimbus, and the following inscription:

Soli sanctissimo sacrum
Tiberius Claudius Felix et
Claudia Helpis et
Tiberius Claudius Alypus filius eorum
Votum solverunt libens (sic) merito
Calbienses de Cohorte tertia.

Lipsius, in the thirty-sixth Op. of his Virgo Hallensis, furnishes a similar form of address to the Virgin Mary as the queen of heaven. "O goddess! thou art the Queen of heaven, of the sea, of the earth, above whom there is nothing but God. Thou Moon, next to him the Sun, whom I implore and invocate; protect and take care of

[21] Quarterly Review. [22] Rev. xii., 1.
[23] Vol. ii., plate 54.

us both in public and private. Thou hast seen us these forty years tossed in a public storm; O Mary, calm this tumultuous sea. Hanc pennam tibi nunc, Diva, merito, consecravit Lipsius." The nations on the borders of the Holy Land paid divine honours to the Moon as the Queen of Heaven; and the Jews suffered themselves to be seduced into the same species of idolatry. The prophet Jeremiah represents them as inviting each other to commit this abomination. "Let us sacrifice to the Queen of Heaven, and pour out our drink offerings to her."[24]

There is a sublime reference attached to the symbol of Seven Stars in the sacred writings of the Jews, that ought not to be overlooked. Stars were sometimes used as emblems of earthly potentates, and at others of the ministers of God's sanctuary; but they have a much higher destination. They represent the Seven Eyes mentioned by Zechariah, which typify the care of divine providence, ever watchful to promote the welfare of his creatures; and the Seven Lamps of the Apocalypse, which symbolize the Holy Spirit of God; whence are also derived the seven spiritual gifts of a Christian man. In the degree of Knights of the East and West, the Seven Stars are explained to signify the seven qualities which ought to distinguish a Freemason, viz., Friendship, Union, Submission, Discretion, Fidelity, Prudence, and Temperance.

The number is remarkable, for it was always considered by the heathen, as well as by the Jews, to designate perfection, and was hence the symbol of heaven. The followers of Mahomet adopted a similar opinion; and the seven editions of the Koran were called by the name of "the seven traditions;" and they believe that the throne of God is surrounded by seven spirits or archangels, called Michael, Gabriel, Lamael, Raphael, Zachariel, Anael, and Oriphiel.

The Holy Spirit of God thus symbolized was known to the pious Jews, for David prays that God will not withdraw his Holy Spirit from him.[25] The streams of water mentioned by Isaiah and other prophets,[26] when the fructifying influence of the Sun should be sevenfold,

[24] Jer. xliv., 17. [25] Ps. li., 11.
[26] Isai. xxx., 25, xliv., 3, &c.

or as the light of seven days, to heal the spiritual wounds of his people, had the same reference. Indeed, "the Hebrew poets, to express happiness, prosperity, and the advancement of kingdoms, make useful images taken from the most striking parts of Nature, from the heavenly bodies, the Sun, Moon, and Stars, shining with increased splendour, and never setting; while calamities, such as the overthrow and destruction of kingdoms, are represented by opposite images."[27] The suffusion of the Holy Spirit, as predicted by the Jewish prophets, is frequently referred to in the New Testament.[28]

Now by the Symbol before us we must understand the one Holy Spirit shining with sevenfold power, as the prophet assures us should be the case when it was communicated to man. The Saviour himself says that "the seven Spirits of God, symbolized by Seven Stars,"[29] are in his possession, and that he will give them to whom he pleases by prayer to his Father;[30] and the Father, on his part, promises to bestow the grace in the name of "the Spirit of his Son."[31] It is described as Seven lamps of fire burning before the throne of God;[32] and they actually appeared on the heads of the Apostles at Pentecost as so many Stars, to represent the accession of light and truth which were then infused, when an universal knowledge was imparted, along with the power of speaking all languages; and no possibility left upon their minds of relapsing into error or misconception while teaching the true system of salvation through a Mediator, and the indispensable necessity, which, from that time forth, was imperative on all people to pay an equal respect to duties, whether to God, their neighbour, or themselves, as a proper preparation for an eternal residence in another and a better world.

[27] Bp. Lowth on Isaiah xxx., 26.
[28] Gal. iii., 14, et passim.
[29] Rev. iii., 1, and v., 6.
[30] John xiv., 16.
[31] Gal. iv., 6.
[32] Rev. i., 4.

LECTURE XII.

Epistle Dedicatory

TO

BRO ALEX. GRANT, ESQ., D. P. G. M. FOR DERRY AND DONEGAL, THE FOUNDER, AND HON. MEMBER OF THE LODGES 126, 265, 279, 282, AND 284 ON THE REGISTRY OF ENGLAND; AND OF 46, 196, 332, 407, AND 589 ON THAT OF IRELAND.

— F. H. M'CAUSLAND,	W. M.
— HENRY S. SKIPTON,	S. W.
— REV. E. M. CLARKE,	J. W.
— J. W. EAMES,	P. M.
— JOHN PRILL,	SEC.
— JOHN KEYS,	TREA.
— WILLIAM THOMPSON,	S. D.
— ISAAC STIRLING,	J. D.

Of the Lodge Light of the North, Londonderry.

MY DEAR BRETHREN,

I beg you will accept my warmest acknowledgments for the distinction which you have been pleased to confer upon me. I am gratified by every new evidence of the effects of my humble labours, because it conveys an unsolicited opinion that they have not been entirely useless. And although I do not entertain the vanity of supposing that the rapid progress which distinguishes Freemasonry at the present day has been produced by any exertions of mine, yet I am not without hope that the course I have pursued for so many years to place the Order before the world in its true position, and to show the connection of general literature with its various subjects of disquisition, has contributed in some slight degree to disarm prejudice, and dispose the initiated to admit our claims to public estimation with somewhat of a better temper than they manifested half a century ago.

Freemasonry is a noble Order, and embraces a fund of information which not only tends to modify the manners and dispositions of mankind in this world, but possesses a direct influence on their preparation for the world beyond the grave. It was the universality of its principles which first enlisted my sympathies in its behalf; and a more extended view of its beauty and usefulness has confirmed the impression, and made it the solace and comfort of my old age.

I have taken the liberty of dedicating the following Lecture on the general import of our glorious symbol, which forms the Consummatum est of Freemasonry, to the W. M., officers and brethren of the lodge, so aptly denominated the Light of the North, because it will display to the inhabitants of the northern districts of Ireland the results of that benign system of Light which we call Freemasonry. It elevates the soul by a graduated ascent to the realms above, founded on that secure basis which is distinguished in Masonry by the peculiar name of Light; and advances the worthy brother from earth to "a celestial canopy sprinkled with golden stars;" thus realizing the expectations of an active and useful life, employed in the duties recommended by the Lectures of Masonry. Past, Present, and Future unite in cementing this delightful consummation. The past is consecrated by memory and Hope; the present by Faith; and the future by Charity; thus completing the cycle of human existence.

But while, as Masons, we thus strive to make our calling and election sure by works of piety and charity, we must never forget that moral virtue alone will not guide us to the summit of the Ladder. The first step is Faith, and on that celestial virtue all our efforts must be based. It is the Great Light which must enlighten our path from the cradle to the grave; and our only safe guide through the devious ways which we are bound to tread in our passage from this world to another. It is the evidence of things not seen, the substance of things hoped for. From this high principle our benevolence should flow in an uninterrupted stream, producing a rich harvest of good works to the glory of our Father which is in heaven.

Such are the doctrines which I believe to be imbedded in the system of Freemasonry, and if they be kept steadily in our view during our mortal pilgrimage, they will gradually advance us step by step on the innumerable rounds of the masonic Ladder, till we attain to that ethereal mansion at its summit where the just exist in perfect bliss to all eternity; where, as our Lectures predicate, we shall be for ever happy with God T G A O T U, being justified by faith in his most precious blood.

Believe me to be,
My dear Brethren and Friends,
Your faithful Servant and Brother,

GEO. OLIVER, D.D.,
Honorary Member of the Lodge.

Scopwick Vicarage,
May 1*st*, 1850.

Lecture the Twelfth.

General import of the Symbol of Glory—the Consummatum est of Freemasonry.

"His birth is as the morning; his strongest time, or his middle time (be his time long or short) is as his noon; and his night is that when he takes leave of the world, and is laid in the grave to sleep with his fathers. This hath been the state of every one since first the world had any on it. The day breaking, the sun ariseth; the sun arising, continues moving; the sun moving, noontide maketh; noontide made, the sun declines; the sun declining, threatens setting; the sun setting, night cometh; and night being come, our life is ended. Thus runs away our time. If He that made the heaven's sun hath set our lives' sun but a small circumference, it will the sooner climb into noon, the sooner fall into night. The morning, noon, and evening—these three conclude our living."

HEXAMERON.

"Elysium shall be thine, the blissful plains
Of utmost earth, where Rhadamanthus reigns.
Joys ever young, unmix'd with pain and fear,
Fill the wide circle of th' eternal year;
Stern winter smiles on that auspicious clime;
The fields are florid with unfading prime.
From the bleak pole no winds inclement blow,
Mould the round hail, or flake the fleecy snow.
But from the breezy deep the blest inhale
The fragrant murmurs of the western gale.
This grace peculiar will the gods afford."

POPE'S HOMER.

THE glorious symbol which forms the subject of the preceding Lectures can be considered in no other light than as a grand and expressive type of the progress of a good and worthy brother from this world to the next. And in that point of view it constitutes one of our most happy emblems, and reflects great credit on the Order. Exclude this comprehensive hieroglyphic, and the Light of Masonry would burn dimly, if not be altogether extinguished. True, there are an abundance of other symbols in the system, which embrace appropriate references, and the meaning of some of them is very significant; but this includes a general view of everything valuable in time and eternity. It commences in the deepest recesses of that celebrated locality which has been received by many sound professors of our faith, as well as by the learned Jewish doctors, as the sacred scene of the last judgment; and terminates in the highest heaven;—it opens in the lowest of valleys, and closes on the holy mountain of the Apocalypse;—it has its origin in darkness, and ends in a burst of glorious light.

Such is the life of man. Generated in darkness, he enters into the world poor, and miserable, and naked. Unable to help himself, he depends entirely on the assistance of others for the preservation of his existence. He sees nothing, he hears nothing, he knows not friend from foe. He is *a point;* a feeble insignificant nonentity, sensible to nothing but mere animal instincts. His life moves in *a circle* of darkness, ignorance, and imbecility; and escaping danger only by the protection of Providence, and the watchful care and attention of those who are his natural guides and guardians, during his helpless, poor, and pennyless state.

This unpropitious view of human nature does not continue long. The initiatory rite of his religion is performed, and his faculties begin to expand. He becomes able to distinguish his friends from strangers; he understands the words of those that are about him; and answers them first by smiling looks, and afterwards by a lisping imitation of words, which soon change into articulate sentences; concise, indeed, but sufficiently expressive to convey the intended meaning. He stands on his feet—he walks—he runs—and the weak and helpless infant becomes a vigorous boy, in the full and happy enjoyment of his newly acquired faculties.

The circle widens. Like a rough ashlar in the hands of the workman, or a lump of clay under the plastic science of the potter, the infant mind becomes moulded into form. He is taught to read, and his intellect begins its work. Thought and reflection spring up as his education advances; and approaching manhood brings him acquainted with the secrets of the Book of Life, where he finds the *two great parallels* who personate faith and practice: by whom he is taught to regulate his life and actions according to their dictates, if he be ambitious of peace in this world or happiness in the next.

At this period of his life, Freemasonry recommends to his most serious contemplation the volume of the Sacred Law; charging him to consider it as the unerring standard of truth and justice, and to regulate his actions by the divine precepts it contains. And he is further told that this First Great Light will teach him the important duties which he owes to God, to his neighbour, and to himself. To God, by never mentioning his name but with that awe and reverence which are due from the creature to the Creator; by imploring his aid on all lawful undertakings, and by looking up to him in every emergency for comfort and support. To his neighbour, by acting with him upon the square; by rendering him every kind office which justice or mercy may require; by relieving his distresses, and soothing his afflictions; and by doing to him as in similar cases you would wish he should do to you. And to himself, by such a prudent and well regulated course of discipline as may best conduce to the preservation of his corporeal and mental faculties in their fullest energy; thereby enabling him to exercise the talents with which he has been favoured by God, as well to his glory, as to the welfare of his fellow creatures.[1]

Such is the recommendation of the two great parallels supporting the circle and point, which is corroborated in the system of Freemasonry, and necessarily include Faith and practice; and having attained these, the candidate is entitled to ascend the first division of the Ladder, through the portal which will be freely opened to him by the gracious Virtue who guards the entrance. In the vigorous stage of manhood, his duties and obligations will materially increase; but if he steadily perseveres in the path chalked out by the Sacred Law of God, he will not find any difficulty in discharging them to the satisfaction of his own conscience. This will afford a reasonable ground of Hope, and enable him to apply confidently for admission to the upper portion of the Ladder. Hope, with a cheerful countenance, opens wide

[1] Dr. Hemming's E. A. P. Charge.

the gate, and the ripened man, animated and enlivened by these two virtues, passes the middle age of life, and his soul ascends slowly, but surely, to the haven of peace, as his weakening body goes downward towards the grave.

Old age succeeds—a time of comfort and satisfaction, after a life spent in the performance of the three great moral and masonic duties. He has no fear of death, because he is prepared for it. The coffin and its mournful embellishments display no terrors to him, because he considers life as the sleep of *darkness*, and death as awakening him from a disagreeable dream to the enjoyment of light and happiness. The peace of God, which passeth all understanding, keeps him firm in the faith; by the aid of which, added to the practice of universal benevolence and love for his fellow creatures, he is enabled to contemplate with calmness and equanimity that event which will separate him from all his earthly friends and connections; because he sees before him, by the eye of faith, a world where everything is bright and glorious; where he shall be reunited to his friends; where sorrow and trouble cannot intrude; and where never-ending pleasures will reward the cares and troubles of his mortal pilgrimage. He approaches the scene of his hopes and wishes with a palpitating heart, and finds the portal of CHARITY thrown open to receive him, and the bodily pains of death are alleviated and cheered by the sound of the angelic host singing the anthems of heaven, and ready to conduct him to that place of rest, where he will wait with patience, in company with the spirits of other just and holy men, till all things are consummated, and the day of resurrection ushers in the eternal reign of the Messiah.

Nothing could be a more wise and just arrangement than the appointment of an intermediate state for the soul, from the time of its departure from the body to the day of judgment. Having been clogged with a corrupt and sinful body, which the Platonists denominated "the bondage of matter," it would scarcely have been in a condition, at the moment of its exodus, to bear either the refulgent glory of God's presence on the one hand, or the extreme punishment of eternal fire on the other. It is true, man is sent into the world with a commission to "go on towards perfection," which, though unattainable in this world, will certainly be completed in the next. And accordingly, while the just are ripening for glory, the wicked degenerate from bad to worse in a similar proportion, as a fit preparation for the perdition that awaits them.

It may be as well to observe here, that this doctrine was embodied in the Spurious Freemasonry; and Olympiodorus, in his commentary on the Gorgias of Plato, thus explains it:—he says, "When Ulysses descended into Hades, he saw, amongst other things, Titius, Sysiphus, and Tantalus. The former was lying supine upon the earth, and a vulture was devouring his liver. The liver signified that he had lived solely according to his animal propensities and the indulgence of his passions. Sysiphus was continually employed in rolling a stone up a hill, which, having attained the summit, escaped from his hands and rolled down again. This was the punishment of ambition and anger; its descent showing the vicious government of himself, and the stone symbolizing the hard, refractory, and rebounding condition of his life. Tantalus lay extended on the borders of a lake, and under a tree bearing abundance of fruit; but he was unable to derive any benefit from either. The fruit which evaded all his attempts implied that he had been living under the dominion of fancy; and his vain attempts to drink out of the lake, showed the delusive and rapidly gliding condition of his life."

In neither of the above cases would the spirit be prepared for a great and sudden change to perfect happiness or perfect misery. The soul of the righteous would not be sufficiently refined and sublimated to endure the blaze of light which proceeds from the throne of the Deity; neither would that of the wicked be fitted to endure the burning wrath of an offended God. Shakspeare alluded to something of this kind when he spake of the spirit of man delighting

To bathe in fiery floods, or to reside
In thrilling regions of thick-ribbed ice;
To be imprisoned in the viewless winds,
And blown with restless violence round about
The pendant world.

On which Douce observes: "with respect to the much

contested and obscure expression of bathing the delighted spirit in fiery floods, Milton appears to have felt less difficulty in its construction than we do at present; for he certainly remembered it when he made Comus say,

> ' * * * One sip of this
> Will bathe the drooping spirits in delight
> Beyond the bliss of dreams.' "

In order, therefore, to prepare the soul for its reunion with an incorruptible body at the resurrection, and to endure the effects of that sentence whose duration shall be everlasting, an intermediate state has been provided by the Divine wisdom and goodness, where the spirit of the just man, liberated from its contact with a material Tabernacle, which obstructed its progress towards the perfection of a future state, receives an acccession of knowledge that is intended to prepare it for final glorification. It floats in liquid ether in a blessed region of light, purified from all gross and sensual appetites and desires, and enjoying a comparative degree of happiness, in a progressive state of preparation for supreme felicity in prospect.

> Beyond the glitt'ring starry sky,
> Far as the eternal hills,
> There in the boundless worlds of Light,
> Our dear Redeemer dwells
> Immortal angels bright and fair
> In countless armies shine;
> At his right hand with golden harps,
> They offer songs divine.
> They brought his chariot from above
> To bear him to his throne;
> Clap'd their triumphant wings and cry'd
> The glorious work is done.

This peaceful abode, or world of spirits, is distinguished in Scripture by the several names of Paradise, Abraham's bosom, the third heaven, and the Hand of God;[2] and it appears that when the soul, which Pope denominates a vital spark of heavenly flame, has shaken off its earthly tabernacle, so called from the Tabernacle of Moses which contained the ethereal Shekinah, it will be conveyed by angels to this peaceful place of rest, there to remain

[2] Luke xxiii., 43, xvi., 22. 2 Cor. xii., 2. Wisdom iii., 1.

until the judgment day. It will be associated with those of Abel, and Noah, and Abraham, and Moses, and David, and other worthy and pious men, who have been admitted into God's *rest*, but not into his *glory*; and will remain in peace, exempted from all pain and disquietude, from all contention and dispute, malice, hatred, and illwill, and secure from the temptations of the devil, until it be God's good pleasure to give them the kingdom. Thus Chrysostom says,[3] "understand what and how great a thing it is for Abraham to sit, and for the Apostle Paul to expect, until they be made perfect, that then they may receive their reward. For until we come, the Father hath foretold them, he will not give them their reward. Art thou grieved because thou shalt not yet receive it? What should Abel do, who overcame so long since, and yet sitteth without his crown? What Noah? and the rest of those times? for behold they expected thee, and expect others after thee. They prevented us in their conflicts, but they shall not prevent us in their crowns, because there is one time appointed to crown all together."

Many curious enquiries might suggest themselves in this place respecting the intermediate state of the soul; as, what is its form? does it assume the figure of one of the five regular bodies? whither does it go? what is its employment? or what its degree of consciousness? Is its place in the air, like that of the evil spirits which frequent "dry places, seeking rest, or go about continually trying whom they may devour?"[4]

"What means these evil spirits use to tempt us we are not distinctly informed; but it is great folly, either on the one hand to doubt the reality of the fact, because we know not the manner, or on the other to entertain groundless imaginations, or believe idle stories, and ascribe more to evil spirits than we have any sufficient cause. For there is no religion in favouring such fancies, or giving credit to such tales; and there has frequently arisen a great deal of hurtful superstition from them. This we are sure of, and it is enough, that neither Satan nor all his angels have power, either to force any one of

[3] Hom 28, in Epist. ad Hebræos.
[4] Matt. xii., 43. 1 Peter v., 8.

us into sin, or to hinder us from repenting, or without God's especial leave to do any one of us the least hurt in any other way. And we have no cause to think that leave to do hurt is ever granted them, but on such extraordinary occasions as are mentioned in Scripture. They are, indeed, often permitted to entice us into sin, as we too often entice one another. But these enticements of evil spirits may be withstood just as effectually, and nearly by just the same methods, as those of evil men."[5] But to return to our subject.

It may be enquired, where are our first parents? Where is Noah, a preacher of righteousness; or the faithful Abraham? The meek Moses; the valiant Joshua; or David, the man after God's own heart? Where are the prophets of the old, or the Apostles of the new Covenant? They are not in heaven, although undoubtedly accepted by the Most High; nor, although favoured with a good report through Faith, have they yet received the promises.[6]

We know from the testimony of Jesus Christ, that the souls of men will possess intelligence, and a knowledge of each other, and of those they have left behind; because it is expressly said that Dives saw and knew Lazarus in Paradise; and therefore it is only fair to presume that the spirits of the departed will recognize each other in the intermediate state. And if they were unable to communicate with their friends in the flesh, Abraham would have told Dives so more explicitly when he requested him to send Lazarus to convert his five brothers. But he said no such thing. He merely replied that it was unnecessary, because they had already the means of salvation in their own hands, if they chose to use them. His words are very remarkable. "If they will not hear Moses and the prophets, neither would they be persuaded though one should rise from the dead."[7]

And we have "a very considerable probability, that St. Paul anticipated on the last day a personal knowledge of those on his part, and a personal reunion with them with whom he had been connected in this life by the ties of pastoral offices and kind affection. That the recog-

[5] Mant and D'Oyley on James iv., 7.

[6] Acts ii., 34, compared with Heb. xi., 39. [7] Luke xvi., 31.

nition would be mutual seems to be a matter of course. And it may, I apprehend, be further assumed, that the same faculty of recognition which would exist at the day of Christ, or at the commencement of the future state of existence, would be perpetuated during its continuance; and that a faculty, which should be allowed to St. Paul and to those with whom he was thus connected, would not be withholden from others who had stood to each other in the same relation, or in other relations of mutual attachment and endearment whilst on earth."[8]

But here another question arises which appears of some importance towards estimating the perfect happiness of the blessed. If earthly friends mutually recognize each other, will not the consciousness that some of them are in torment materially allay that happiness by the existence of sensations of regret and sorrow for their fate? This argument has been often urged by deists and infidels as an insuperable objection to a future state of rewards and punishments. Be it the province of Freemasonry to refute it.

The spirit, when it departs from its earthly tabernacle, becomes purified from all gross and carnal affections. Faith and Hope are consummated and extinguished; and nothing remains in the glorified state but Charity or universal love. Impure passions or feelings can have no existence there; and as the minds of the condemned must have been essentially and wholly vicious, no predilection in their behalf can possibly remain in a spirit which has been cleansed from all its earthly thoughts and feelings; for such a reflection would imply a doubt of the divine justice, which would be sinful; and the spirits of the just are incapable of sinning. Besides, the ties by which we are united on earth, even of husband and wife, parent and child, or brother to brother, are weak and feeble compared with the bond of perfect love which cements the angelic society of another life. They neither marry nor are given in marriage, but become the children of God, and exempt from all the infirmities of their former imperfect state. The friendship of the wicked is forgotten, and every accepted soul inherits a

[8] Bp. Mant's Happiness of the Blessed, p. 79.

fulness of joy and eternal blessedness, in a society where all bodies are glorified, and where the perfect faculties are incapable of sorrow or regret.

An extended speculation on these points, however, for which inspiration furnishes no certain guide, might lead us into error. Some, indeed, think that the spirits of our departed friends are our guardian angels; that they are continually with us;—wherever we go, they follow us—grieving when we do wrong, and rejoicing when we do right. Others think they are employed by the Almighty as angelic messengers, to distribute mercy and loving kindness to other worlds.

Now supposing that the spirits of our departed friends should be thus employed—and it is not improbable—the doctrine affords a valuable lesson of patience under bereavement; and shows the folly of grieving for their loss, as if there were no hope for them. We are subject to pain, and sorrow, and distress; but they are exempt from all such feelings. They have nothing but happiness, and peace, and joy. If, therefore, they are appointed to watch over us; to preserve us from harm, and to guide us in the ways of truth and virtue, we have greater occasion to rejoice than to grieve, for their loss is better for us as well as for them. In fact, there can be no doubt of their happiness after death if they have done their duty here. They have been sown in corruption, but they will be raised in incorruption; and the natural body which has been deposited in the earth will be converted into a spiritual body.

Again, we are ignorant of the satisfaction of having all our wants supplied, and all our wishes gratified; and therefore we can form no conception of the state in which we should be placed after death. But we may be quite certain that if we perform with undeviating punctuality our respective duties to God, our neighbour, and ourselves, as they are exemplified in the lectures of Masonry, as well as in the Holy Book which crowns the Pedestal, we shall certainly partake of the happiness which is reserved for all faithful Brothers, although we cannot perfectly understand it.

The situation of this place of rest has not been revealed, and therefore all conjecture would be inadequate to discover it. Whether it be above or below the earth the

Scriptures do not say, and it would be rash to pronounce an opinion on such an abstruse subject. In knowing that it is the entrance to the haven of eternal rest, or final salvation, we know enough to satisfy any reasonable enquiry. We are told, indeed, that Enoch, Elijah, and Christ *ascended*, and St. Paul was caught up into Paradise;[9] and therefore we may reasonably presume that it lies beyond the sphere of the remotest stars; but in what region or situation we are perfectly ignorant. St. Paul calls it the third heaven; which, according to Macknight, is the seat of God and of the holy angels, into which Christ ascended after his resurrection; but which is not the object of men's senses as the other heavens are; the first being the region of the air, where birds fly; and the second that part of space which contains the stars.

In that place of rest will the souls of just men remain, in the enjoyment of each other's society, and the interchange of those amenities which we cannot at present comprehend, but which we are assured will constitute supreme felicity; clogged with none of those vile or boisterous passions, and bereft of the distressing wants and necessities which encumber our earthly body. Here will be no need of laws, because there is neither property to protect, wants to supply, or necessities to provide for. Labour will be in no request, for the soul is impalpable, and requires neither food nor raiment. Locks and bolts to guard against intrusion will be useless, for in that holy place, moth will not corrupt nor thieves break in and steal. Warlike weapons will be unknown in a region of universal love and peace, where "the wolf shall dwell with the lamb, and the leopard lie down with the kid; and the calf and the young lion and the fatling together, and a little child shall lead them. And the cow and the bear shall feed; their young ones shall lie down together; and the lion shall eat straw like the ox. And the sucking child shall play on the hole of the asp, and the weaned child shall put his hand on the cockatrice's den. They shall not hurt nor destroy in all my holy mountain: for the earth shall be full of the knowledge of the Lord, as the waters cover the sea."[10]

[9] Heb. xi., 5. 2 Kings ii., 1. Acts i., 9. 2 Cor. xii., 4.
[10] Isaiah xi., 6-9.

This state of comparative felicity will be enjoyed by the spirits of all good and worthy brethren, until the number of the elect is completed,[11] and the sound of the eternal trumpet shall announce the day of judgment, and reunite the soul to its old companion the body, in a more glorified state, for "flesh and blood cannot inherit the kingdom of God, neither doth corruption inherit incorruption."[12] This change affords the best mitigation of the concern so apt to overwhelm us on account either of our own death, or of the death of those who are very dear to us. The bodies of the righteous are not swallowed up by the grave, as a prey, but deposited there as a trust; which will surely be demanded back again, and of which a punctual restitution will be expected. So our admirable Liturgy has taught us, in the office of interment, to commit the bodies of the deceased to the ground, in sure and certain hope of the resurrection of the righteous to eternal life, through our Lord Jesus Christ; who, as we are assured by St. Paul,[13] shall change our vile body—this miserable earth, and ashes, and dust—that it may be like unto his glorious body, according to the mighty working whereby he is able to subdue all things to himself.[14]

The general resurrection has been embodied in the third degree of Masonry; and the reward of duty is pointed out in our glorious Symbol. That is the great harvest when the wheat and the tares, the worthy Mason and the unsainted cowan shall be gathered together for final separation.[15] And it is a remarkable coincidence, that in the system of Freemasonry the very same symbols are used to illustrate the same facts and doctrines as in Christianity. Thus, at the solemn consecration of our lodges, we use *corn* as the emblem of perpetuity and immortality, because the vital principle is never extinguished. It will keep for thousands of years without the germ of vegetation being injured or destroyed. And even when deposited in the ground, although in appearance it dies, and crumbles into dust and corruption, it springs into a renewed life, and bears fruit thirty, sixty, and a hundred fold.

[11] See the Church Burial Service. [12] 1 Cor. xv., 50.
[13] Phil. iii., 21. [14] Stanhope's Com. on the Epistles, as above.
[15] Matt. xiii., 39.

The Jews entertained a similar notion respecting the human body; and believed that after death it contained a certain indestructible part called *luz*, which is the seed from which it is to be reproduced. It is described as a bone, shaped like an almond, and having its place at the end of the vertebræ. This bone, according to the Rabbis, can neither be broken by any force of man, nor consumed by fire, nor dissolved by water; and they tell us that the fact was proved before the Emperor Adrian. In his presence, Rabbi Joshua ben Chauma produced a *luz;* it was ground between two millstones, but it came out as whole as it had been put in. They threw it into the fire, and it was found to be incombustible. They cast it into the water, and it could not be softened. Lastly, they hammered it on an anvil, and both the hammer and the anvil were broken without affecting the *luz*. The Rabbinical writers support this notion by a verse from the Psalms, "he keepeth all his bones; not one of them is broken." A dew is to descend upon the earth, preparatory to the resurrection, and to quicken into life and growth these seeds of the dead.[16]

The First Great Light explains the nature of the last judgment by the symbol of corn growing in a field; and the process is gradually unfolded from the sowing of the seed to the gathering of the produce. And its coincidence with certain masonic ceremonies, to which I have just referred, will render its illustration acceptable at the close of these lectures. The subject is of such importance, that T G A O T U himself thought proper to give his hearers a particular description of it. He opens the august subject by comparing the kingdom of heaven to corn in the seed which a man sowed in his field. The seed itself was the emblem of man; he who sowed the seed was the Most High at the creation; and the field represented the earth. The enemy, full of mischief, sowed tares among the wheat. No sooner was the world called into being, than the evil spirit entered into the garden of Eden, where our first parents lived in perfect happiness, and endeavoured to persuade them to rebel against their Creator. Unfortunately they complied with his request; and, as we learn from the old Royal Arch

[16] Quarterly Review, vol. 21.

Lecture, brought misery upon themselves, and us, and all mankind. Thus were tares sown among the wheat; or, in other words, the cowan and the Mason became mingled together in the world, although it is impossible they can ever meet in the lodge.

But as the tares could not be distinguished until the corn was sprung up, so the consequences of their sin were not fully exemplified till Cain murdered his brother; an event which it is extremely probable originated some of our most occult ceremonies. And if we were to trace the progress of error from then to the present time, we should distinctly understand the extent of the injury which had been inflicted upon mankind by the enemy who sowed tares amongst the wheat; which forms a valid reason why cowans are so carefully excluded from our private assemblies; for it will easily be seen that, as the wheat represents good and worthy Masons, they are symbolized by the tares.

It will not be inapplicable to our present purpose to consider further how each class will be dealt with at the harvest, when the Judge shall appear in the Cloudy Canopy, attended by his holy angels in the valley of Jehoshaphat. At this period the wheat and the tares, the good and the bad, must inevitably appear. They have had the same means and incentives to the practice of their social and religious duties; and now they are to be disposed of according as they have used or abused them. Every person that has ever lived in the world will be present; whether the Great Assize be really held in the valley of Jehoshaphat or elsewhere. Kings, princes, and prelates; masters, wardens, and brethren; the expert architect, and the humble artisan, all must obey the summons. Multitudes from every quarter of the globe, however distinguished by colour, nation, or language, will be assembled. The call is universal; penetrating to the utmost extent of this capacious lodge; from north to south, from east to west, from surface to centre, from earth to heaven.

But how many will tremble for fear? Will the wicked —the cowans—try to hide themselves? Where will they go? Into the caves of the rocks which abound in that celebrated valley? They will find no protection there, for the rocks will be broken in pieces at his presence

Will they flee to the east or to the west—to the utter most parts of the earth or sea? There he will find them out. Every attempt at concealment will be as ineffective as that of the hunted ostrich who buries her head in the sand.

It behoves every brother, therefore, to consider how he will be able to bear the investigations of that day when God will bring every work into judgment, with every secret thing both good and evil; and how, as a Free and Accepted Mason, he has improved the advantages conferred by his initiation. Have the Lodge Lectures performed their office effectually, and produced the fruits of piety to God and good will to man? Have they caused the unmetallic key to hang as Masonry requires? Have they cemented the masonic chain, and produced a reciprocation of fraternal benefits? He has possessed superior advantages; and where much has been given much will certainly be required.

It is an awful question to consider, how we shall bear to have our actions, our wishes, our very thoughts exposed in the presence of all our brethren. I am afraid it would overwhelm the very best of us, if we duly reflect on the awful position we shall then occupy under the penetrating Eye of the Great Architect of the Universe, seated in a canopy of clouds, and surrounded by the angelic host. The pious Bishop Hall says, "if the law were given with such majesty and terror on Mount Sinai, how shall it be required at the last day? If such were the proclamation of God's statutes, what shall the judgment be? I see, and tremble at the resemblance. The trumpet of the angel called unto the one; the voice of an archangel, the trumpet of God, shall summon us to the other. To the one, Moses, that climbed up the hill, and alone saw it, says, God came with ten thousands of his saints; in the other, thousand thousands shall minister to him, and ten thousand times ten thousand shall stand before him. In the one, Mount Sinai only was on a flame; all the world shall be so in the other. In the one there were fire, smoke, thunder, and lightning; in the other, a fiery stream shall issue from him, wherewith the heavens shall be dissolved, and the elements shall melt away with a noise. O God, how powerful art thou to inflict vengeance upon sinners, who didst thus forbid sin! And if thou wert so terrible a Lawgiver, what a Judge shalt thou appear!"

The Great Architect of the Universe being thus seated on the Cloudy Canopy, attended by the hierarchy of heaven: every eye will be fixed upon him, and amidst the most profound silence the Books will be opened where the actions of all mankind have been registered by the finger of God. And from their evidence the whole human race will be separated into two distinct classes; viz., the faithful brother will occupy the one, and the obtrusive cowan the other. The former will be approved, and placed in the north-east, on the right-hand side of the Judge, as successful candidates for his mercy; and it will be observed that a tradition has universally prevailed, that He will come in the East, and be seated on a Cloudy Canopy facing the West; while the latter will be placed on his left hand as candidates rejected.

The Judge will then proceed to pass sentence on both which can never be reversed. He will declare his approbation of those good and worthy brethren who stand at his right hand; applauding their Faith, their Hope, and their Charity; and give them immediate possession of that holy place which is veiled in clouds and darkness beyond the summit of the Ladder. And he will declare the reason why they are thus distinguished. It is because they have accomplished those moral and religious duties which are recommended in the Lectures of Masonry. They have fed the hungry, clothed the naked, visited the sick, and relieved the distressed. Surprised as well as gratified at this public communication of his divine will and pleasure, because they entertained reasonable doubts of their own unworthiness, they tremblingly ask—"When saw we thee hungry, or thirsty, or naked, or sick, or in prison, and ministered unto thee?" He answers the question and confirms the sentence by saying,—"Verily I say unto you, *inasmuch as ye have done it unto one of the least of these my brethren, ye have done it unto me.* Come, ye blessed of my Father, enter into the joy of your Lord!"

While this is going on, what are the feelings of the irreclaimable cowans who have been placed on the left hand; who have been disobedient to the laws, or corrupt panders to heterodoxy; whose proselyting zeal has converted saints into sinners? They have had neither Faith nor Charity, and are now bereft of Hope. They feel that the time of repentance has passed away like a vision of the night, and the hour of punishment is at hand. They regret that they have neglected all opportunities of improving their talent; but their regret is unavailing, for it comes too late. The time is past. The gates of Faith, and Hope, and Charity, are closely tyled, and they cannot gain admission. They have had time to repent, and have not repented;—they have had calls out of number to reform, and they have not been converted. T G A O T U, therefore, after condemning their falsehood, their profanity, and their unrepented sins, pronounced the final sentence which consigns them to everlasting punishment.

He then graciously proceeds to justify the sentence. It is because they have been reprobate and profane—atheists and unbelievers. They have neither fed the hungry, clothed the naked, visited and relieved the sick and comfortless, nor performed any of the common duties of humanity. And therefore he consigns them to that place of darkness and despair prepared for the devil and his angels; while his faithful followers are transferred by the angelic host, who are in attendance for that very purpose, from their exalted situation in the north-east, to the Grand Lodge above, where they will exist for ever in perfect charity and perfect happiness.

LECTURE XIII.

Epistle Dedicatory

TO

BRO.	R. P. HUNT,	W. M.
—	G. SOUTHALL,	P. M.
—	W. H. FLETCHER,	S. W.
—	EDWARD DAVIS,	J. W.
—	S. PRUCE,	TREA.
—	EDWARD DAVIS,	SEC.
—	EDWARD HAMMOND,	S. D.
—	W. TAYLOR,	J. D.
—	T. BAKER,	J. G.

Of the Lodge Hope and Charity, No. 523, *Kidderminster.*

DEAR FRIENDS AND BRETHREN,

My labours are drawing towards a conclusion, and the time approaches when it will become incumbent on me to retire from the Craft, and take a grateful leave of the fraternity by whom I have been uniformly treated in the most kind and distinguished manner. During a period of nearly half a century since my initiation, and of forty years' active exertion to promote the general interests of the Craft, I have persevered, amidst evil report and good report, in my endeavours to place Freemasonry before the public as a moral and scientific institution which is eminently calculated to produce the universal happiness of mankind. And I believe it will be found that throughout all my numerous publications, there is not a page which is at variance with the benignant principles of the Order; as I am sure I never intentionally penned a single sentence to wound the feelings or excite the wrath of any individual brother. Even when I have found it necessary to vindicate myself from calumnious attacks, I have invariably endeavoured to preserve a respectful tone towards my accusers, and am not conscious of having ever exceeded the bounds of a temperate and graceful style of controversy. My aim has always been, in conformity with a well known passage in the Lodge Lectures, to speak as well of a brother in his absence as I would have done had he been present; and when that could not be done with propriety, I have adopted the Mason's peculiar virtue—Silence.

For this reason, amongst others, I have been honoured with the patronage of the noblest and best of men and Masons; amongst whom I am proud to include the two illustrious princes, the Dukes of York and Sussex; the Archbishops of Canterbury and York; the Duke of Leinster; the Earls of Zetland, Yarborough (late), and Aboyne; two Bishops of the diocese where I reside; Sir Edw. Ffrench Bromhead, Bart.; Richard Ellison, Esq.; and many other distinguished personages in various parts of the globe. The patrons of this my final work, which constitutes the cope stone and crown of my masonic publications, are the brethren of those lodges by which I have been more particularly distinguished; and the fraternity at large, wheresoever dispersed under the wide and lofty canopy of heaven. To this supernal abode it is hoped that every true and worthy brother, who has been fortified by Temperance, Fortitude, Prudence, and Justice, and has passed up the Ladder through the gates of Faith, Hope, and Charity, will eventually arrive.

The name of your Lodge includes a description of the blessed process which, by steady perseverance, will lead to those happy mansions where the just exist in perfect bliss to all eternity; where they will be for ever happy with God, the Great Geometrician of the Universe, whose only Son died for us that we might be justified through Faith in his most precious blood. This is our Hope, that we may all finally meet in that blessed abode of never-failing Charity and it has constituted the animating principle which has supported me through all the arduous trials of an eventful life; and still forms the sincere and only wish of him who has the honour of dedicating his closing Lecture to you, and to subscribe himself,

Dear Friends and Brothers,
Your most faithful and obedient Servant,
In the holy bond of Masonry,

GEO. OLIVER, D.D.,
Honorary Member of the Lodge.

SCOPWICK VICARAGE,
June 1, 1850

Lecture the Thirteenth.

A Recapitulation, or general Summary of the doctrines contained in the preceding Lectures, with their application to the system of Freemasonry.

"English Masonry is the knowledge of the eternal God, as the God of Creation and Providence; it is also the knowledge of the Lord Jesus Christ, as the God of redemption; and far from ascribing creation to a concuitous adhesion of matter, we believe in a God who created all things; far from ascribing the wonderful mysteries of Providence to the blindness of fate and chance, we believe in a God ordering all things both in heaven and earth; and in all the steps of masonic advancement we cry, Hosanna to the Son of God! Blessed is he who cometh in the name of the Lord! Hosanna in the Highest!"

INWOOD.

"Then shall appear the sign of the Son of Man in heaven."
MATT. xxiv., 30.

It may be expected, as this volume constitutes the completion of the plan which I had formed when I first undertook the responsibility of entering on a virgin soil, and turning up a glebe which had scarcely been touched by the hand of man, that I should wind up my labours by a brief analysis of the general design of the treatise, as a work which is especially devoted to the purpose of explaining the tendency and final consummation of the Order.

The Book is intended to be a type of the masonic institution. It opens with a view of the present state of the science, considered as a means of producing spiritual perfection. On this point I am anxious to avoid any misinterpretation. Freemasonry cannot accomplish this result single handed, but as contributing its aid in connection with other agencies. No one can become a Mason without a sincere profession of a belief in One God, the

Great Architect or Creator of the Universe; nor can he give his assent to our ordinary Lectures without an application of the types of the Old Testament to the manifestations declared in the Gospel; or in other words, without an acknowledgment of the truth of Christianity. If Christ be not the Messiah predicted by the Jewish prophets, then the Lectures of Freemasonry are nothing more than an agreeble fiction; pleasing, perhaps, to the fancy, but without carrying conviction to the judgment; and consequently, useless as a stimulus to moral duty, without which the summit of the Ladder can never be attained, or the portal of Charity opened. And the remarkable coincidences which I have brought into one focus, will show that all the princpal truths of revealed religion have been concentrated in the Lectures of Freemasonry. They, who think otherwise, are not only deceived themselves, but are deceivers of others.

The present state of Freemasonry is distinguished by its numerous charitable institutions; which, by removing the attention from the affairs and disquietude of this world, leave the worthy brother at leisure to prepare for another and a better. Its application to the sciences is not so obvious. And as its founders, in the early and mediæval ages, were archæologists and ecclesiastical architects, it appears reasonable that, in addition to its moral reference, some marked attention should be paid to those pursuits which distinguished our ancient brethren, and produced that eminence which made their example worthy of imitation in the establishment and perpetuity of an institution founded expressly on the arts which raised them to distinction, and made them exemplars of every Christian virtue.

For this purpose a revision of the Lectures has been suggested as eminently calculated to restore Freemasonry to its primitive purity and usefulness; and to implant in the minds of the brethren a veneration for all that is great and good; inciting them to emulate those glorious examples of morality, combined with the sublimities of science, which have enrolled our ancient brethren in the lists of the benefactors of mankind. It might appear invidious to select individuals from the catalogue of this noble band of Masons, as being worthy of peculiar note; but I cannot refrain from holding up to the notice of the fraternity, as deserving of commendation, a few eminent Masons whose names are familiarly known in this island; such as Lanfranc and Gundulph, William of Sens, W. Anglus, John of Gloucester, Irwin von Steinbach, Nicolas Walton, Robert de Skellington, Geoffery Fitzpeter, William of Wykeham, Henry Chichely, Wolsey and Cromwell, Denham, Vanburg, Sheldon, Jones, Wren, Web, Sayer, Desaguliers, Anderson, Dunckerley, and many others, who were all expert Master Masons in different ages, and their fame will never die.

The Lodge Lectures have been repeatedly arranged and rearranged, to keep pace with the progress of human enlightenment; and each revision has been an improvement on its predecessor, and tended to increase the popularity and standing of the Order. The present period requires extensive alterations and additions; for as the Lectures form the real touchstone by which a true judgment may be formed of the application of Freemasonry to the requirements of an improved state of society, they ought to approximate as nearly to perfection as possible. Thirty-six years have now elapsed since the last revision; and when it is considered what rapid strides have been made during that period in the education of the people, and how extensively knowledge has been propagated, it is time our leaders began to consider the consequences of resting on their oars, while their contemporaries are pulling with might and main that they may be the first to reach the goal.

From these considerations, I have suggested a plan in my First Lecture by which the united wisdom of the Craft might be brought into requisition, to remodel our ordinary Lodge Lectures on such a principle as may prove acceptable to the members, and contribute to the best interests of the Craft.

As the object of the preliminary Lectures is to place Freemasonry on its true basis, and to exhibit it as an institution capable of promoting human happiness in this world, and inspiring the hope of attaining to a more perfect felicity in the next, the Second Lecture is devoted to a dissertation on the poetry and philosophy of the Order, for the purpose of showing that any attempt to become an adept in its mysteries is sure to fail, unless it be the result of a scientific research into the hidden meaning of our signs and symbols, where all our secret lore has been deposited, and in which our occult doctrines can alone be found. It has been truly said, that whatever a man most loves will constitute the poetry of his life, and the philosophy of his soul. It encourages him to admire things unknown, till admiration is turned into reality; and thus he conquers the difficulties which appeared to impede his progress to knowledge; he acquires a mastery over wonders, which distance had magnified into sublimity, and makes apparent impossibilities yield to the all powerful force of industry and perseverance.

Having considered the present flourishing condition of the Order, springing out of a proper understanding of its poetry and philosophy, I have proceeded, in the next place, to take a more particular view of the Lodge Lectures, as the expositors of Masonry, because its very existence depends on their adaptation to its genuine principles and practice. They consist equally of science and morals; and strongly recommend the practice of the moral and social duties of life, as a passport to the Cloudy Canopy, which is attainable by means of a Ladder, whose principle rounds or staves are Faith, Hope, and Charity.

These Lectures can only be attained by persevering industry and sedulous application, for there is no such thing as an intuitive acquisition of science, as Knittel, the Jesuit, pretended. The formula being neither written nor printed, there is no alternative but to acquire a knowledge of it by oral communication from the lips of the W. M. in open lodge. As Euclid said to Alexander the Great, "There is no royal road to Geometry," so say I to the anxious Mason. And his endeavours to acquire this knowledge will be rewarded by other advantages.

If industry and regularity be systematically followed, they will soon become habitual, and tend to the prosperity of all his worldly pursuits. *Aide toi et le ciel t'aidera.* Such was the advice of Jupiter to the clown whose cart wheel was sunk in a slough so deep that his horse was unable to extricate it. The fellow sat him down quietly on a bank, and cried out, "O, Jupiter, help me!" "Help you!" said the god. "Lay your shoulder to the wheel, you lazy hound, and endeavour to help yourself, and then you may expect assistance from me." Industry and application are therefore recommended in the Lecture under our consideration, as the only means of becoming acquainted with the details of Freemasonry, and acquiring the reputation of being, what our transatlantic brethren aptly denominate, "a bright Mason."

The English fraternity is divided into two parties, both powerful from intellect and position; one of which is impressed with a conviction that Masonry will be extended and ennobled by an open promulgation of those doctrines and practices which are peculiar, but not necessarily secret; while the other section adopts the creed of those "scrupulous brethren" of the last century, who committed many valuable documents to the flames, lest they should fall into the hands of Dr. Anderson when he compiled the original Book of Constitutions by command of the Grand Lodge. These would have Freemasonry to be a stationary institution, depending solely on the faith of oral tradition; and hence they decry all disquisitions which possess a tendency to increase its influence or improve its details. I have examined the peculiar opinions of these two sections of the fraternity in detail; and it is presumed that a decision has been impartially pronounced according to their respective merits.

In the Fourth Lecture we advance an important step in our investigation, by showing that the doctrines enunciated in the Lodge Lectures are consonant with the teaching of our holy religion, as its morality is explained in the inimitable Sermon on the Mount. And although Freemasonry is not a religious sect, yet it inculcates the duties which belong to every religion "in which all men agree." This constitutes the great mistake, equally of those who are ignorant of our mysteries, and of those also who have only a superficial knowledge of them. They are apt to fall into the error of taking an extreme view of the subject, and pronouncing either for one alternative or the other; either Masonry is a system of infidelity and excludes religion altogether from its disquisitions, or it is a religious sect which would supersede the necessity of Christianity, and monopolize the office of procuring, unaided, the salvation of man.

The truth, however, lies between these two propositions. Freemasonry is neither an exclusive system of religion, nor does it tolerate the detestable principles of infidelity. It is a teacher of morality, and contributes its powerful aid, in that capacity, to the salvation of souls, by recommending and enforcing the duties of the second table, and demanding an acquiescence in the doctrines of the first. And this course of discipline is perfectly consonant with the teaching of Christianity. When the lawyer asked the subtle question, "Which is the great commandment of the law? Jesus said unto him, Thou shalt love the Lord thy God with all thy heart, and with all thy soul, and with all thy mind: and thy neighbour as thyself;" or in other words, this is all that is required by the Jewish law for the salvation of man.

In the Fifth Lecture we enter *in medias res*, by shewing how we deduce occult doctrines from visible symbols; and for this purpose we commence with the most obscure emblem in Masonry, and one which has taxed the ingenuity of the Craft ever since its introduction into the Lectures. Conjecture has been very busy about the circle, point, and parallel lines; and the consequence is that all have arrived at the same end, although they have traversed different paths to attain it. The explanation of the symbol has varied, but the doctrine which has been deduced from it remains pretty nearly the same; and the reader is furnished with a detail of the several versions which have prevailed at different periods, and under the sanction of successive Grand Masters.

And this will be an answer to those who think Freemasonry should remain stationary, and never deviate from the position which it occupied at its first establishment in ages far remote; but in the universal movement of Nature and Art towards that perfection which the Great Creator originally designed to bless his creatures here, as a humble taste of the glory which will be revealed hereafter, Masonry alone ought to be quiescent, and deprived of those benefits which the improvements of an enlightened age have conferred on all other institutions. It was an ancient rule of conduct given by Musonius, a heathen, that "those who are desirous of improving their morals must be continually employed in amending and reforming their lives by the improvements of philosophy." If, therefore, the fraternity refuse to profit by the advances which science and art are gradually making in every successive age, they will be worse than the heathen, who had no better guide than the light of Nature to direct their enquiries amidst the dense ignorance by which they were surrounded.

If the intelligent portion of the Craft in every age of its existence had been thus bigotted, it would long ago have succumbed to the pressure on all sides, which has been arrayed against it, and have been forced out of its place like some foreign substance that had been unnaturally introduced into the human frame. But the different interpretations of the circle, point, and parallel lines, assure us that our rulers have ever considered Freemasonry to be progressive, like all other sciences, and have profited by the light which has from time to time been thrown upon it, to improve its details, and render its doctrines the pride of the fraternity, and the envy of the world.

The Sixth Lecture is exclusively devoted to a consideration of the doctrines embodied in this expressive symbol. The greatest error which we discover in its interpretation, is that which confines the Deity to the centre of the circle. He is present every where; and were it possible he should withdraw the light of his countenance from any part of the universe, how small or insignificant soever it might be, not only would that locality be thrown into irretrievable confusion, but as the order and regularity of each part is essential to the support of the whole, the balance would be destroyed, and the vast fabric of Nature instantly dissolved.

> The least confusion but in one, not all
> That system only, but the whole must fall.
> Let earth unbalanc'd from her orbit fly,
> Planets and suns run lawless thro' the sky;
> Let ruling angels from their spheres be hurl'd,
> Being on being wreck'd, and world on world;
> Heav'n's whole foundation to the centre nod.
> And Nature trembles to the Throne of God.
>
> Pope.

But if the Deity were confined to the centre, he would be absent from every other part of the Universe, and thus the doctrine, that the vast machine is upheld solely by the power and providence of God, would be justly questioned, and the infidel would exult in an imaginary triumph. But it is not so. Freemasonry affords no grounds for the triumph of infidelity. It is founded on the knowledge and acknowledgment of God the Creator, who fills all space, extends through all extent, and is every where present to hear and answer the prayers of his faithful worshippers.

The two perpendicular parallel lines have also been variously interpreted; some understanding them to refer to the Tabernacle and Temple of the Jews, as represented by their builders, Moses and Solomon; while others interpret them to be symbols of the two St. Johns. But either view of the case will terminate in an application to Faith and Practice. For Moses, according to his own evidence, was a type of Christ, whom he speaks of as a prophet like himself, and, therefore, was an object of Faith to the Jews, as St. John is to Christians; while Solomon, who carried out the incipient idea of Moses in the construction of the Temple, was a personification of that *practical* religion which St. John the Evangelist recommended so powerfully to his followers, as the perfection and fruit of Faith. If, therefore, a candidate for the honours of Masonry represent the central point of Time, as it is now understood, and his circumambulation be indicative of his progress to eternity, the perpendicular parallel lines can be no other than the Faith and Practice by which he expects to attain the object of his Hope, those supernal regions of universal Love which will endure through everlasting ages.

And these, like the Ladder of Masonry, must be based on the Three Great Lights; which accordingly form subjects of disquisition for the Seventh Lecture. The Holy Bible is the proper object of a Mason's Faith; by the Square he learns to modulate his Practice by its dictates, conformably to the rules of morality and justice, and the Compasses instruct him to limit his desires within the boundary lines of reason and revelation, which constitute the two limbs of that conprehensive instrument, as the only certain method of avoiding the evils which always attend the unrestrained indulgence of the passions.

The Ladder of Masonry, with its innumerable rounds or staves, which constitutes the steep ascent from earth to heaven, and connects them together by the sacred Tracing Board at its foot, and the Cloudy Canopy at its summit, forms the next portion of our Symbol to be examined. Its principal steps or Gates are three, corresponding with the three Theological Virtues, which are the leading characteristics of every candidate for a residence above the Cloudy Canopy, being, indeed, the graduated stages of a Christian's life.

In the Symbol I have ventured to pourtray this Canopy, and such attendant emblems of the secret places wihch it conceals from mortal view, as may be clearly drawn from sources of undoubted authenticity; for I should have considered it highly indecorous to indulge in any flights of fancy, or unauthorized visions of the imagination, while dealing with such a profound and sacred theme. The legitimate symbols of the Deity, as they exist in our own Scriptures and the writings of the early fathers of the Christian Church, as well as in the symbolism of the more ancient dispensation, are the equilateral triangle, the rainbow, the hand, the vesica piscis, and the All-seeing Eye; and these I have surrounded with the hierarchies of heaven praising God and saying, "Holy, holy, holy, Lord God Almighty, which was, and is, and is to come!"

Each of these Symbols is fully, and it is hoped, satisfactorily explained in the Ninth Lecture, for the purpose of showing that the Free and Accepted Mason who has performed his duties faithfully, and discharged his obligations truly, may be presumed to have made a successful progress up the masonic Ladder—to have fought the good fight of Faith, to have been supported by the anchor of Hope, and to have practised Charity in both its divisions, by a cheerful and active benevolence on the one hand, and by kindness, and good will, and brotherly love to all mankind on the other. Thus, as his pilgrim age draws gradually towards its termination, he beholds by faith, a prospect of the blessedness which is reserved for his enjoyment when the gates of death are closed upon him, and the fiat of the Judge shall place him at his right hand.

Now, the application of these principles to Freemasonry leads to an unerring result which forms the subject of the concluding Lectures. The candidate commences his probation by the acknowledgment of the being of a God, on whom his hope and trust are firmly placed.

He knows that difficulty and danger attend his Christian course, and he presumes that his masonic progress may have a similar tendency. He is told that all his trials may be surmounted by Faith and Hope. If he believes a thing impossible, his despondency may make it so, but if he perseveres to the end, he will finally overcome all difficulties. This consideration constitutes an effectual shield against infidelity. If he believes in God, he must also believe him to be a present help in times of difficulty and distress. If he be omnipresent he is also omnipotent; and wherever faith is firm and trusting, it is sure to meet with all necessary assistance.

As this blessedness of a future state is indicated by the Blazing Star, which points to Christ, Jehovah, or the Son of God, and the salvation which was wrought out for mankind by his death upon the Cross, I have endeavoured to ascertain not only the true meaning of the Symbol, but also the reasons which induced our ancient brethren to introduce it into Masonry. In a moral sense it was formerly called Beauty, and referred to Prudence; and spiritually, because the Divinity, in the ancient hieroglyphics, was always designated by a Star, it represents the Star in the East which guided the Eastern Magi to Bethlehem to worship the Great Jehovah—Him who came down from heaven to take away the sins of the world, and to teach mankind the way to blessedness by the exercise of THREE pre-eminent virtues which form a constituent part of the system of Freemasonry, "the greatest of which" is placed at the summit of the Ladder that leads to the Grand Lodge above.

Here, then, we see the triumph of Freemasonry. It aims at an eternal residence in the skies, of which our Cloudy Canopy is a significant symbol, accessible by Faith, Hope, and Charity, based on the Holy Bible, supported by an altar decorated with the Circle, Point, and Parallel lines, and situated on Holy Ground in the consecrated Valley of Jehoshaphat.

And what is the Ladder that leads to the Throne of the Inaccessible?

PRAYER.

POPULAR WORKS ON FREEMASONRY

PUBLISHED BY THE

MASONIC PUBLISHING AND MANUFACTURING CO.

432 BROOME STREET, N. Y.

☞ **Any book in this list sent by mail to any address in the United States, free of postage, on receipt of the price.**

A CYCLOPEDIA OF FREEMASONRY; containing Definitions of the Technical Terms used by the Fraternity. With an account of the rise and progress of Freemasonry and its Kindred Associations—ancient and modern: embracing OLIVER'S DICTIONARY OF SYMBOLICAL MASONRY. Edited by ROBERT MACOY, 33d. *Illustrated with numerous Engravings.* Cloth, gilt side, ... $4 00

GUIDE TO THE ROYAL ARCH CHAPTER; a complete Monitor for Royal Arch Masonry. With full instructions in the degrees of Mark Master, Past Master, Most Excellent Master, and Royal Arch, according to the text of the Manual of the Chapter. By JOHN SHEVILLE, P. G. H. P., of New Jersey, and JAS. L. GOULD, D. G. H. P., of Connecticut. Together with a Historical Introduction, Explanatory Notes, and Critical Emendations. To which are added Monitorial Instructions in the Holy Order of High-Priesthood in Royal Arch Masonry, with the Ceremonies of the Order. By JAS. L. GOULD, M. A., 33d. Cloth—gilt back and side............................... 1 50

THE MASONIC HARMONIA; a Collection of Music, Original and Selected, for the use of the Masonic Fraternity. By HENRY STEPHEN CUTLER, Doctor in Music, Director of the Cecilian Choir, etc. Published under the auspices of St. Cecile Lodge, No. 568, City of New York. Half-bound—cloth sides, $1 00...........per doz. 10 00

THE GENERAL AHIMAN REZON AND FREEMASON'S GUIDE, containing Monitorial Instructions in the Degrees of Entered Apprentice, Fellow-Craft, and Master Mason, with Explanatory Notes, Emendations, and Lectures: together with the Ceremonies of Consecration and Dedication of New Lodges, Installation of Grand and Subordinate Officers, Laying Foundation Stones, Dedication of Masonic Halls, Grand Visitations, Burial Services, Regulations for Processions, Masonic Calendar, etc. To which are added a RITUAL for a LODGE OF SORROW and the Ceremonies of Consecrating Masonic Cemeteries; also an Appendix, with the Forms of Masonic Documents, Masonic Trials, etc. By DANIEL SICKELS, 33d. Embellished with nearly 300 Engravings and Portrait of the Author.

Bound in fine Cloth—extra—large 12mo.................................. 2 00

" " Morocco, full gilt, for the W. Master's table, with appropriate insignia of the East.................................. 3 50

THE HISTORICAL LANDMARKS and other Evidences of Freemasonry, explained in a series of Practical Lectures, with copious Notes. By GEORGE OLIVER, D. D. 2 vols. Large duodecimo—with Portrait of the Author. Cloth, $5 00. Half Morocco................. 7 00

WASHINGTON AND HIS MASONIC COMPEERS. By SIDNEY HAYDEN, Past Master of Rural Amity Lodge, No. 70, Pennsylvania. Illustrated with a copy of a Masonic Portrait of Washington, *painted from life*, never before published, and numerous other engravings. Cloth—Uniform Style, $2 50. Cloth—full gilt—gilt edges, $3 50. Morocco—full gilt.. 5 00

www.ingramcontent.com/pod-product-compliance
Lightning Source LLC
Chambersburg PA
CBHW020502270326
41926CB00008B/703

* 9 7 8 3 3 3 7 3 9 6 4 0 4 *